I AM THE ALIEN

CLYDE R. SMITH

I AM THE ALIEN

I AM THE ALIEN Copyright © 2022 by CLYDE R. SMITH. All rights reserved.
No part of this publication may be reproduced, stored in a retrieval system, or transmitted in any way by any means, electronic, mechanical, photocopy, recording, or otherwise without the prior permission of the author except as provided by USA copyright law.
The opinions expressed by the author are not necessarily those of Book writer corner.
Published by Book Writer Corner
433 Walnut Ct Pittsburgh PA 15237, USA
www.bookwritercorner.com
At Book Writer Corner, we have established an authority ghostwriting service provider. Having been served a plethora of clients ever since our inception in the industry, we have developed into a formidable force of writers, editors, proofreaders, etc… Our primary objective is to facilitate clients to complete their e-Books and other related projects without any problem. Writing a book or pursuing any other gig, especially when you are already overwhelmed with loads of office work, could be quite stressful. The same aspect forces you to completely withdraw from writing your own book. However, at Book Writer Corner, we extend our support and encourage you to carry on despite being loaded with other responsibilities. We can help you achieve your dream of writing a book.

PREFACE

Going over a small hill and down into a valley that borders the river that I was taking one nice sunny morning led me to this very unusual tale of an adventure of a life time for this young man.

This is not my story, but I must tell you how this tale came into my possession. As I was walking along just enjoying the warmth of the day, a glint of light reflected off a piece of metal caught my eye. Being a bit of a curious person and with a small detour from my path, this took me to a fairly large canister of some kind. One side was ripped open with small pieces of tree bark protruding along the metal. It had apparently hit the tree as it crashed to the ground.

Bending down to get a closer look at whatever this thing was gave me quite a start to see a partially decomposed body lying inside at a lop-sided angle. The odor was overwhelming from rotting flesh, but with only myself in the area, the thought of not inspecting

Further never crossed my mind. There would be plenty of time to report it to the authorities on my trek back home.

I didn't want to touch the body as the stench was revolting to my sense of smell as it was, so I didn't want to disturb it further. In trying to see what else was inside I spotted a folder with a large stack of papers. By stretching my arm as far as it would reach, the folder was close enough to be extracted from the capsule.

Standing there for a while thinking about what to do with the whole situation, it seemed the logical thing was to go home, call the police, and let them take care of it. The task of finding out who he was, and if any of his family was still around would be up to them. But that folder of papers was going home with me to see what was in them. I would give the authorities the papers if it seemed that it would be useful to them, but certainly, after I had looked at them first.

When I returned home, what I found written there was a fantastic story of what had happened to this person that was in that crashed capsule. So here is his story of what happened, how he got to another planet, and the amazing life he lived there.

I AM THE ALIEN

    My name is Daniel J. Woolsey. This is a record of what happened on how I got to a planet far, far way, and of my life on that planet. I was just a typical young college student struggling to make decent grades. I wasn't smart enough, nor enough of an athlete to get a scholarship but did good enough on the entrance exams to be accepted. Wanting to participate in the events on a collage campus, I became a "walk-on" to try out with the wrestling team. Then the unthinkable happened.

Not being a good scholar in English and spelling, this account probably will not be very accurate in punctuation or spelling. Sometimes it will be in the past tense as I remember back on how it started, but now that I have progressed so far, it will be recorded as I live it. Sometimes it will be a conversation taking place as it happened, and maybe sometimes as I remember it when I have time to write. Then at times, it will be interspersed with my thoughts on the situations that come up, and how I dealt with them.

I pray that someday this record is found by a person from the planet Earth, or at least by some civilization that can figure out how to read English. I will begin with how it started with that first encounter. So here is my story.

With my back pressed against the tree, and legs straight out in the cool grass that October afternoon, I was feeling a bit guilty at having skipped out on wrestling practice. The class work was completed, and would not be due until Monday at the University.

I was tired from the week's rigorous practices, and the weight training was getting to me. So this warm sunny afternoon, my longing to get back to the enjoyment of simply going out for a walk with my 22 rifles cradled in my left arm would do much to recharge my internal batteries. Even though I carried my 22, there would be no need to shoot anything. It just reminded me that I could if I wanted to, and a reminder of how good a shot I was as a kid. Leaning the rifle against the big old oak tree, and stretching the legs out in the grass was a perfect position to just let the mind wander and daydream about nothing in particular.

Getting tired of the sitting position it felt good to stand again. Deciding to head over to the next hill would take me to a favorite spot to hunt squirrels. But with no intentions of actually shooting one as this would lead to gutting it, and then skinning it when I got back, and I sure didn't feel like cooking it today. So I left my rifle leaning up against the tree and headed for the hill. I would pick it up on the way back.

There was a nice flat grassy glen at the downside of this hill, then up and over the next hill would take me to a stream that was so clean and clear you could take a big cool drink to quench a thirst, and not worry one bit about how clean and safe it would be to drink that water.

Cresting the hill and beginning the trek downward the sight that stopped me in my tracks was totally unexpected. It had been several months since walking over the crest of this hill, but I couldn't believe how fast somebody had gotten this building set up way out here between these two hills. It must have been a prefabricated one that was brought in a piece at a time, or else those carpenters were really fast. Looked like a lot of work still needed to be done, as there were no windows cut into the shell from what could be seen from here.

I didn't know who owned this part of the valley, and it sure seemed odd that they would put up a building way out here where it would be difficult to get to. Might as well check it out anyway as I was going right by it on my way to the stream. Even as a kid that was a good adventure hiking over these hills to get there. Leaning over the edge of the stream to stick my lips down into the cool water seemed a bit risky not knowing exactly what might be in that water so this made it a bit more exciting to my imagination. The tip of my nose always had to go into the cool river water in order for my lips to suck in a nice drink. That was a magical moment to wipe off the nose on my shirt sleeve as I savored the cool water sliding down the throat.

But I never made it to the river. Getting close to the building it was starting to look like a giant plastic igloo. I had seen smaller versions of fiber glass or plastic ones in back yards, or for demonstrations, but nothing the size of this one. Not wanting to intrude kept me at a convenient distance, but as curiosity got the better of me, and no one seemed to be around, it seemed innocent enough to get a closer look.

The entry at the front was open, and a ramp leading up to that opening was too inviting to pass up the opportunity to sneak a peak inside. A good knock on the side of the entrance brought no response, so maybe just a quick step inside would satisfy my curiosity, and I could get out of there before the builders or the owner would get back.

Stepping through the opening there was a funny feeling sensation in the air, and it seemed like I was hearing something, but couldn't make out what such a faint sound might be. The inside was really strange, with equipment fastened to the walls that made no sense for any purpose I could imagine.

The sound of what seemed like a bear lumbering up the ramp instantly pumped enough adrenaline into my veins to spur the "fight or flight" action. My gun was still back leaning against a tree, and I wasn't about to take on a bear with my bare hands. Grabbing a door handle behind me I quickly jumped in pulling the door shut behind me. Standing there with one hand still on the door handle staying as quiet as nerves would let me, maybe that bear would head on out after sniffing around for a few minutes.

Suddenly embarrassment and guilt flooded my brain at realizing my mistake. The workers or carpenters must have returned. The sounds were slightly muffled and sounded like they were speaking Spanish because I couldn't understand a word of it. How could I explain my presence, and justify getting caught in a closet? Sooner or later I would have to come out and face up to it. Taking a few moments to ponder how to make it amusing about being scared of a bear, maybe they would laugh at the situation, and just let me walk out without even a scolding.

Then the sound of an engine winding up to speed, with a slight vibration coming into that dark closet stopped me from exiting the hiding place. With feet frozen in place, moving was impossible. I just plain couldn't move. Why I don't know, but feet wouldn't work. Hands didn't work, and what was more terrifying, the brain also seemed paralyzed as a shudder ran through the building. Thrown to the floor by a force of gravity greater than any force I could overcome, all senses went blank.

Finally, things started working again, so it was necessary to get on facing what was happening, with those workers, and continue on my way. Opening the door my whole body went stiff as standing in front of me was a vision I had always considered non-existent except in the minds of kooks, or mentally unbalanced people.

He stepped back in surprise at seeing me but immediately recovered to call out to others for help. More of them (I think it was four) hurried to help. Pulling me through the doorway, arms were pinned to my side and held there with some sort of restraints.

I was bigger than any of them, and physically could have thrown them all sideways to make a run for the exit, but it was too late, as we were already in the air and accelerating rapidly. The ensuing chatter made absolutely no sense to me, as nothing they said sounded like any language I had heard before. Listening to the talk I guessed they must be debating what to do with me, so took the opportunity to look more closely to see if any features were recognizable. It only took a quick look to see that there was no resemblance to any people on this earth.

I am six feet two inches tall, so they must be about 5 feet to around 5 feet, three inches tall. Slight of build with a head that seemed a little too big for the body. The eyes also appeared too big for the head. Looking into those eyes was like looking into a small, pitch-black tunnel with no end.

Standing there together while this vessel picked up speed it seemed nobody knew what to do. Restrained by the four of them, they appeared to be terrified by finding such a big intruder in their midst who was just as scared of what was going to happen now. The acceleration of the ship smoothed out, so still holding onto my arms they pulled and pushed me close to a wall where a restraint went around my left ankle. With their captive now securely fastened to the wall, they stepped back and began chattering away. I assumed all the talk was about what to do with me.

Two of them headed for the front. A short burst of talk to a superior officer who returned with them to stand there staring at me. Then with very little discussion, I could see what had been decided, as the movement of hands, and body actions made it obvious that being killed, and ejected out into space was to be my inglorious fate. Thank God there was one objector. So I still had a few moments to live.

After some conversation, the objector hurried away, and in a few moments came back with another one of them. A thorough examination began by this new arrival. Eyes, ears, nose, and throat was given consideration. Peering inside my mouth again, he looked inside the mouth of one of my captors, and back again at me. Motioning with his hands at my shirt and buttons, it didn't take a genius to know what he wanted. So with the shirt off, his hands probed here and there, squeezing my muscles. Poking the pectorals his head turned to look at the others with a comment that seemed to bring a laugh or some sort of amusement. I could see from their slight build that any comparison of muscle would put them on the losing end of the bulk measurement.

I tried to keep a pleasant look and smile to look as harmless and friendly as I could be even though I was very afraid of what was to become of me. With more discussion, all of them left to tend to other duties except the one doing the inspection. So the conclusion on my part was that he must be the group's doctor. The doctor continued with his inspection. Feet, hands, range of motion, and apparently anything he could think of to satisfy his curiosity continued until he was satisfied.

To keep him interested, and to delay the time of ejection into space. So as he looked at each part of my anatomy I named that part. When he picked up one of my shoes for inspection, I said shoe, pointed to it and repeated, "Shoe". Darned if he didn't repeat "shoe" every time. So instantly I shook my head yes, smiled, and repeated the word. He got the idea and pointed to my bare foot sticking out in front of me. I immediately said foot several times. Finally, he repeated the word "foot" and named it in his language. Immediately repeating his word as I pointed to the foot brought a look of satisfaction from him. Taking the initiative, my finger went from my eye, ear, mouth, etc. naming each part with him repeating it after me, and then my repeating his names as well. I struggled to remember each of his pronunciations but wondered if this was only slowing down the ultimate execution.

The doctor seemed to have his curiosity satisfied, turned to leave, and headed to the front of the ship for a discussion with the one I supposed was the Captain, or whatever they called the one in charge.

I sat down on the floor for what seemed like several hours. Fatigue was beginning to take over so lay down for what seemed even more hours. One by one the inhabitants of the ship came by to stop and stare for a few moments before going on about whatever it was they did on this ship. The doctor woke me up to put a cushion on the floor to lay on. I did eventually doze off again for some needed sleep. After all it would not be right to send a tired man to his death I sarcastically reasoned. Then if on cue the doctor returned again with a container of food which brought another wry smile of "can't send a man to his death that was hungry". I didn't know what it was called, but the taste was pretty good so I smiled and expressed my thanks to this kind doctor.

After a long discussion between the doctor and the captain, they both came back to where I was still tethered to the wall. The doctor continued pointing out body parts and pieces of clothing with both of us repeating the names in both English and then in his language. Moving back a short distance their discussion continued for a few moments.

The captain went back to his command post. The doctor removed my tether and motioned for me to follow. Where could I run to now? Why would I try to do damage to anyone, or to the equipment? The same thoughts had to be going through their minds as well to set me free from the restraint of the tether. So dutifully following the doctor led to a tour of the ship's sleeping quarters, feeding area, and sanitary area.

With no idea of where we were going, or what would be my fate after we got there my nerves were shattered. Going from fear to excitement of this new experience, the problem of communication, what must I learn; how do I learn; what is the food; or, am I the food; what about clean clothing; all of these things kept my mind too busy wondering about it that there was no hope of getting the nerves settled down any time soon.

By the time our tour was over, I had seen most of the ship, and most of the inhabitants had seen me. The feeling of being stared at by each and every one of them must be the way a very attractive female must feel when every man stares at her as she walks down a busy street. It was very uncomfortable, and yet made me feel empowered as well seeing so much attention focused on me. As we returned to the starting point, the doctor made motions of putting something into his mouth so I concluded we would be eating something soon.

We followed a group of eight into the section where tables and chairs were arranged for all of us to sit together. Covered small trays were already in place. When we were all sitting a signal was given, and each one removed the top part of their tray. Utensils were on one side area, with food in two separate compartments of the tray. I watched the ones directly across from me, and pretty much copied their actions. It didn't make any difference what was being eaten because I was definitely very hungry. The food had been processed into a mashed potato consistency to facilitate storage. There was no conversation or comments from anybody. Whether it was because of my presence they remained silence or military-like protocol, but I sure wasn't going to say anything.

When everybody was finished, the alien at the end of the table pushed a button, and we all stood up. Another button was pushed, and the area the trays were sitting on tipped downward where each tray slid out of sight where they probably went to a machine that cleaned and sterilized everything. I assumed the others headed back to their job to make way for the next group coming in to eat.

The doctor motioned for me to follow him again. This time it led to another compartment with bunks lining a wall on opposite sides, and an ally way between. Each bunk had a curved glass cover that slid down over the side to make a closed in bed for sleeping. A few more steps and there were several bunks with sleeping aliens. They looked very comfortable lying there in complete silence. Each bunk had a symbol, with some smaller symbols just below, which I took to be a number, so each one must have an assigned bed.

Going back to the first bunk, the glass cover was rolled up. The doctor motioned for me to get in. I dutifully slid out of my shoes, and got in. There was not enough head room to sit up straight, so leaned back and stretched out on the bed. As I turned my head to smile at the doctor and indicate this was very comfortable, the glass lid came down. I must have looked like panic set in so the glass immediately went back up. This made me feel a little foolish, so patting the bed and smiling I hoped he would make out that it was intended to tell him how comfortable it would be to sleep in this bed.   He reached up behind my head, withdrew, a headset, placed it over my ears, and turned on something. The glass cover came down again. A slight hissing sound started. At first, the sound was like a tape deck getting ready to play music, then I noticed a slight movement of air around the bed, and my body felt very light as though it could almost start floating. This was very relaxing, and as the eyelids went shut, breathing slowed considerably and I didn't object as sleep came to stay.

When I awoke after what seemed like only a few moments, the mind felt refreshed, but for some reason, the body felt stiff. Turning my head to the side to look at the doctor, he was not there. In his place was somebody I hadn't seen before. Opening the glass side the legs would not cooperate as the attempt was made to swing them over the edge in order to get up.

'Easy now," the alien said. "Move slow and easy for a while. The mobility will come back to normal soon. You have been gone for much longer than you know."

I couldn't believe it. It was not English being spoken, but I could kind of understand what he was saying. Seeing the puzzled expression, it was explained how the ear phones was gradually teaching some of the language being used as I slept. There was a lot to be learned yet as this was only a start.

As he helped me move around the compartment to regain some mobility it was impossible for me to determine how long the period of suspended animation lasted, as my time based on earth days was not the same at their time. This new doctor explained that everybody on that ship went through several periods of being suspended that made the trip seem a lot shorter and sustained life for the long trip home. "We don't know how long your system will tolerate suspension so you may have to be monitored closely for some time to see what will work for you," was what he explained.

Talking the best that I could with this doctor gave me an opportunity to get a closer look at him. As I said earlier he was about six to eight inches shorter than me. The head was smooth and bare of hair. A protective ridge of hair was above his two eyes that was considerably thicker than a human's eyebrows. A similar ridge of hair was over the ears on each side of his head. The neck was a little shorter than on a human. The mouth was pretty close to the same shape as mine, though a little smaller. But what was really strange was the nose. He had two; one below each eye. What did he need two for? I couldn't help but wonder what he would do if he got a nose bleed in both at the same time. Most everything else was two of each, so why not a nose? Both noses were a lot smaller than a human's, and maybe between the two of them, they could determine the direction of an odor similar to the way ears can determine sound direction. Likewise, the body was built pretty much like a human, though lighter in build. I couldn't see what his feet looked like, but the hands had four fingers, with what could be called a thumb on both sides of the hand. This ought to be a hand that could grasp most any shape or size of an object.

I must be quite a sight to them. Here they are with no hair on the head or anywhere else I could see. And me with a set of short cropped head of hair, and whisker stubble all over my chin. As though to reassure myself, my hand went to the top of my head to discover I had a good head of hair that was longer than when I climbed into that container. Same with the whiskers on my chin. They had grown considerably. Nothing I could do about that, so figured I might as well just wait to see if survival was even going to be possible with this situation. It was very difficult to grasp what might be going to happen when this ship reached its home land. With no idea of what their planet would look like; what the culture might be; what plants or animals might inhabit the land; what living conditions may be encountered; and especially the temperament and how friendly the population might be to an alien. So I decided to learn what I could about these people as we traveled through space.

After some time (I had no idea of how much time was passing, as there were no reference points for comparison) I learned that when my walk out into the timber on earth took me to where the ship was located the whole group that wasn't sleeping had taken containers to the stream to renew their water supply. When I entered the ship an alarm was sounded that I couldn't hear but got their attention to hurry back to see what the alarm was about. I was surprised that they didn't just kill me or throw me off the ship. But the ship had taken off before they found me, so maybe couldn't open the door once they were in the air.

I wondered how much time had passed on earth, and how much time had passed before they missed me at the university as well as at home.

Somebody was sure to have found my rifle leaning up against that tree. There would be a search for my body. I felt sorry for my family that would be out looking and waiting for answers to the disappearance, and what happened. But I couldn't dwell on that, so was determined to make the most of my captivity.

The next few days (I assumed the time was about a day in length) I tried to learn as much as possible and practiced talking to whoever would help in getting the sound right. This language had sounds so different it was difficult to understand, but I kept trying. There were questions on the culture, and what was happening on their planet. My crude attempts at the language was tolerated with amusement, but no scorn was shown at my effort. They wouldn't let me touch any of the controls or any of the apparatus that made anything work. So I was assigned to keeping things clean which was an experience in itself. The part that was the most surprising was how everybody cleaned themselves and put on clean clothes.

They simply removed all clothing, walked into a small enclosed compartment, and pushed a button. A machine similar to an ultra-violate light rotated around the body. This apparently sterilized, or neutralized any and all impurities. If the hands were really dirty or grimy they would be put through an opening that scrubbed them clean with no effort on your part.

A cover went over the head, and then from other nozzles a spray of spun material encapsulated the whole body. This was followed by another strange light that dried the material into a body-fitting, but loose material. It was surprising how loosely it fit, and how comfortable it was by the time I walked out of that "shower and dressing room". The one thing I wanted to ask was "how do you get out of this thing to use the bathroom?" but this became clear when one of them simply reached down to separate the material with an easy tug with the fingers. When finished, just place the material back together again, and it will stay. Seeing my puzzled expression, it was quickly explained that on a trip like this, strong material was not necessary to operate the ship, so disposed and reconstituted, easy-to-use material was not only economical but quick and easy. If we need stronger, more durable uniforms, they are folded, and put into the storage room to use when we go outside the ship. This was followed by a demonstration of how to separate the material from where needed, and then how to fit it back together again.

My learning was coming along very nicely now, as more of them were willing to talk with me and show how things worked. That sleep session with the headphones was a good start and worked well enough to help get me started with the assimilation of their language and culture. These people turned out to be very helpful, and willing to share information that I asked for, or obviously needed to know.

They were very curious about me as well. My whiskers had been growing for a while now, and there were no supplies to be used for shaving. Most of the ones at one time or another had felt my beard, and then the hair on my head. This seemed to be a great curiosity for them. So in turn I would rub my hand over their bald shiny smooth skull. I figured that was only fair. If they could feel and tug on my hair, then I could do the same to them, though there was no hair to feel. They seemed to be very curious about my skin color as it was touched carefully with a finger and rubbed as if to see if any color came off on it. I sometimes wanted to do the same thing to them as they were all a much darker shade than me. There didn't seem to be any variation in their skin color which seemed a little strange to me. How could it be that there was no variation in skin color? Were they all so genetically alike that the gene pool was limited to only one shade of color?

I had been very curious why they had stopped on our planet. The doctor explained that the water supply on the ship had been recycled many times. With each time it was recycled there would always be a little lost in condensation. When their instruments located a river that was pretty clean of filth and bacteria they stopped to resupply the ship. The clearing was easy to land, and it was a short trip up over that hill to reach the water. There didn't seem to be anything in that area, so guards were not posted. The alarm brought them back. Most of the water containers were now full from that stop. We should have a sufficient supply to finish this trip I was told by one of the engineers.

Being inside with video screens to show what was outside was interesting as we blasted through space. I had no idea where we were going, or what suns and planets were passed on the way, but I could see that the crew had a myriad of charts, and symbols that were being followed. I had hoped for a lesson on reading and following the charts, but they seemed too busy keeping an eye on guidance though space to bother with me. So I just looked over their shoulder from time to time, and at the screen and marveled at how they were keeping track of it all.

After what seemed a very long time alternating between cleaning duty, lunch time, sanitation time, and practicing language skills, I was checked by another doctor, escorted to the glass containment bunk, and put to sleep for another period of time in hibernation. This time I was aware of the how and why of it all. The main reason being that it would suspend my growing older as it took such a long time traveling through space. So with a rotation of one group in hibernation, and two groups alternating work duty and rest periods, the three groups could go long distances.

I went through these cycles two more times, and each time I learned a little more about these aliens. Getting ready for another sleep period they informed me this would be the last time for the hibernation period. That everybody on the ship's roster would be awake and functioning normally when we arrived on our home planet.

This time when when the sleep period ended, the only one still in the compartment was myself. The doctor helped me out, and get adjusted to moving again. When it seemed to him everything was back to normal, he went with me to the front of the ship where the only one left on board was the captain, myself, and the doctor. The doctor quickly explained why everybody else was not there.

"The council had to be informed of your presence, and how you came to be on board. So you were left in hibernation for three days while it was determined what to do with you," the doctor explained. "We will go up the front, where the captain has something to say to you," the doctor continued as he led the way.

"We have discussed your situation with the council. Your qualifications were given careful consideration. There were only two things to consider. You would either be put to death, or a chance to see if a life in this society could be useful to this culture, and to yourself," the captain laid it all on the line.

Didn't seem to be much of a choice for me so there was no hesitation to reply, and I did so with the proper attitude, as I learned early on that it payed to show respect.

"If it pleases the council I would like to offer my services any way that can be helpful to the people of my new home. Please convey my thanks for the opportunity to do so," was my reply.

"That will be be sent to the council immediately. If the reply is positive then you will be taken to a home on the outskirts of the city where you will live with an elderly couple. They will teach you the ways of our people, and the customs of our land. If you are successful in adjusting then you will be free to go where you wish. If not successful then as you were informed of the other choice it will be up to me to carry out the sentence, as I was the one responsible for bringing you here." This was said in a matter-of-fact manner that left no doubt it would be carried out if deemed necessary.

The captain went to the ship's communication center. Called in my response and waited for the answer which came back in short order.

"Get your old clothes on, you are going to the country," was the only short comment the Captain had for me.

I actually didn't know whether to be happy or just choose death and get it over with. All my friends, family, and every memory of my life were in the past. Was this how an immigrant felt coming to the United States? Not knowing anybody; not even able to speak the new language; leaving your family and friends behind, and for what. At least they would be with humans, and look about the same with about the same needs. It went through my mind that suicide could always be one way out if it all became too uncomfortable. But the desire to live is strong, so I must at least give it a try, and see if adjusting to a new life will work out okay with time.

With the old clothes on I said my thanks to the doctor and the captain. Standing outside the ship it was my first chance to look at it a little more carefully than when it was first spotted in the glen back on earth. I wondered now how I could have not seen what it was then, and just ran back up the hill to watch from a distance.

If I ever get back to Earth, the people will want to know what life was like with these aliens, so I would do my best to give them a written record of what happened to me, and about their culture. The plants and animals were probably genetically different, so trying to keep a record of everything that was seen, and done. An Earth camera and computer would be nice to record it all, but so far I don't even have a pencil or paper.

When I get to my final destination, I will do my best to leave a record somewhere somehow. It will probably be a mix of what happened in real-time with my recollection of it.

A vehicle pulled up beside us. Two aliens got off of it to guide me onto a seat. The vehicle started forward without a sound, or at least no sound I could hear. The two beside me and even the driver had been briefed on who, and why they were escorting me to my new home.

The translation between their language, and English is very difficult for me, and since Earth is the only place I know of that has people that can read, it is best if this is all written in English as it will do these people no good to read what they already know about this planet and themselves.

Therefore I will start out by referring to this vehicle as a car: a car that makes no noise. The top was about seven feet wide. The open-side concept reminds me of an old-time Model T touring car from the 1920"s. The silence rules out combustion, and the wide flat top means it must be sun-powered electric. It moved along at a decent speed, and I estimated it took about fifteen minutes to arrive at a house.

As we were getting out of the car, a man and a woman came out of the house. There was no explanation needed since they had already been advised of what I was doing there. This was the first time I had seen any kind of a smile from the time of my capture, and up until now. Filled with trepidation I tried my best to show a pleasant friendly face. I had no idea of the protocol on how a greeting was done, but I stuck out my hand and smiled. The man took it in his, and let me guide it slightly up and down in greeting. The woman in turn did likewise, and I was invited into their house. The car turned around and left without saying a word.

They seemed as curious about me as I was about them. They had been told about my being somewhat lingual, so occasionally repeated a comment using simpler words. We got along great at that first meeting, and spent much of the time asking each other about life on our respective planets.

At that time there were so many questions I wanted to ask, but rather than overdo it, that in time and by having patience, everything would be answered.

Soon it became time for sleep. The man (whose name in their language I still can't pronounce properly) led me to a bedroom. I will just call him Fred for now. It was just starting to get dark, but I noticed it was rapidly getting darker. Either this planet was spinning rapidly, or something was blocking the sun. I would ask tomorrow. Fred pointed out the "sanitation" room, the location of the sleep wear, the used clothing I could wear, and the temperature adjustment dial, and guessing my next question, explained that when they were heard stirring it would be time to get up.

The sanitizer worked the same way as the one on the ship, so as soon as Fred left I stepped in, pushed the button marked "clean" and felt the warm glow as the light did its job on the skin bacteria. There was no spray fabric nozzles here, so having regular manufactured clothing again was a welcome improvement over how that spray stuff felt.

Unfolding the night clothing that was laying on the bed, it turned out to be just a big old night shirt type. I really preferred pajamas, but sure wasn't going to quibble over using a night shirt. Getting into bed, the thought of the ship forced me to look at what was at the front and behind my head to make sure nothing would put me into hibernation. It took some time with thoughts churning around on how tomorrow would go, but eventually got to sleep.

The next morning also came too soon. It seemed this was an awfully short night, so asking a question on the length of a day and night was going to be taken care of before breakfast, if there was even going to be breakfast. I had not had a good plate of eggs, bacon, toast, etc. for lord knows how long now, and the thought of that made me ache for a good old breakfast like we get on Earth. Both Fred and Wilma were up and around (since I called the man Fred, what came to mind was Fred and Wilma Flintstone cartoons, so Wilma is what I would call the female.) I had no knowledge of how long they had been waiting for me to rise and shine. I will have to learn the correct pronunciation of their names soon. As it would probably show a positive attitude toward my assimilation into their culture.

The morning meal consisted of what looked like Brussels sprouts, but tasted more like an unseasoned meat ball. There was a round flat pancake-sized grayish-colored something that had a pleasant taste, and odor. It was an okay breakfast. Since I didn't get a chance to ask before we ate, this seemed like a good time to ask some questions on this planet.

I learned this was a small planet that rotated on its axis somewhat faster than the speed of Earth. Hence the days and nights were much shorter which led to periods of sleep and working hours passing by about twice as fast. I would estimate a day's time would be close to eighteen hours when compared to twenty-four hours on Earth. In time my system would adjust to going to sleep faster, and sleeping very sound for the short period of time it was night on this planet. Their clock told time just as ours does, but adjusted to their planet's system. Once the time divisions were understood it was no problem reading their clock. It sure was different not using seconds, minutes, or hours, and I had to eliminate those thoughts, otherwise getting the units of time heard here mixed in with it, would temporarily confuse my brain.

I learned there were two continents separated by an ocean on each side. Each continent had its own government. There was a competition held each year that determined a king that had some small power over both continents. Both continents had their own sports, but neither continent would compete with the other. The explanation as to why, is that it would lead to nationalistic feelings, then controversies follow, and in the past that led to war. There was so much destruction from the last war that the leaders finally came together, and banned war of any kind, as well as all other competitions between the continents except one that would be just a symbolic gesture. That would be an annual fight between just one people from each place. This was held in alternate countries. There would be a festival designed to satisfy nationalism and keep peace in the land. The winner would be the hero for his country for one year. A long tournament leading up to the next year's match would start, and any feelings of superiority would be abated by the local matches in each district.

That car was an interesting ride yesterday to get there, and I was curious in the propulsion system. I knew it was solar but asked about different sources of power. There is no power used from fossil fuels. Apparently, no subsurface pools of oil, so no refineries to make gasoline in enough quantities to use in combustion engines. So how did you get enough power for the interstellar flight was the logical next question for Fred?

That is a whole different sort of power. They had figured out a way to separate material into matter and anti-matter. Fred didn't know how the anti-matter was kept separate; just knew that small atoms at a time was fed into a combustion chamber. The ensuing explosion produced tremendous power. I will try to learn more lately.

Fred was getting tired of all the questions so took me outside to explain what some of my duties would be in helping with the work during my time living with them. The work would not be hard, mostly for something to do while learning about the culture, and how to get along with the inhabitants.

Going around to the back of the house there was a pen with sides about six or seven feet high with close-knit mesh for the sides as well as the top. Inside the pen were the strangest-looking creatures I've ever seen. Another name I don't know how to write in English, so I'll call it a blob for now because that's about what it looked like.

The color was a mottled gray. It had a body shaped like an extremely fat goose. A short fat neck with a round baseball size head on top. Did not appear to be any eyes, or beak, or anything that looked like a mouth; just a round gray ball on top of the neck. I didn't see any feet or legs so they must have been squatting down on top of their legs.

Fred showed me into a nearby shed, picked up a bucket, and dipped it into a container to fill it about half full as he explained this was enough to provide a whole day's ration for this pen of blobs. Entering the pen he scattered the finely ground feed around the floor of the pen. The feed looked similar to coffee you would use in a percolator.

I had to ask.

"How do they eat that feed? I don't see any mouth or a place to eat it."

"They don't eat the way we do"' Fred replied. "These aren't animals the way you think of animals. These things are fungus. They come down on top of the feed, and absorb it though the spores on the bottom part of their body."

"But if they are not animals, how do they move to get over the feed" I inquired with such a puzzled look that Fred had to laugh.

"Keep watching and wait awhile. You will see how they do it," was his answer'

So I stood there waiting for something to happen. Fred stood there with me and kept right on smiling at the thought of what I was going to see. All of a sudden without any warning, a blob shot about four feet into the air and came down a short distance from where it had been sitting. It happened so quickly that the surprise made me jump backward a bit.

Turning to look at Fred, he was roaring with laughter at my reaction to the blob's jump and continued to chortle while he explained.

"You couldn't see the appendage when it was sitting on it. The blob sits on whatever it is using for food. The fungus absorbs the nutrients along with moisture. Most of the moisture is filtered into the part that is connected to the appendage. The sun warms this area enough that the moisture slowly evaporates swelling that part. When the pressure is high enough it is released all at once slamming the appendage down on the ground which propels the blob upward. It is fairly light weight. The greater the pressure compared to the weight determines how high it goes. When it lands on a new spot it usually lands on some new food for it to use," and Fred finally stopped laughing.

I had to know, so I had to ask. "How do you get the next crop of little blobs?"

"These are fungus. So when they reach maturity, they are put in a separate pen where they just sit and spores spew out all around them. The spores are scraped up and scattered on top of food so it can be absorbed. That's how they are started. Then as they grow the feed is modified for maximize growth. This is a very nutritious food. You had some this morning. You will have more of it mixed in with other things later on today," and Fred moved on as he finished this explanation.

Going around to the other side of the pen was a regular garden-type plot of ground with vegetable-type plants of root crops, greens, berries, and things. We walked through several rows of crops with Fred explaining their use, how it was prepared, what is was called, and the special care some needed for maximum production. There were different kinds of fertilizer used to feed different groups of plants, so he would take care of that as we went along.

The one plant that was the most interesting was what looked like a tree but was only about twenty or so feet high. All around the perimeter were vines I supposed but looked more like a braided rope hanging down. About five feet from the ground were things that looked like purple egg plants. This would be a staple in the diet for most of the people on this planet. It was explained the flavor would vary depending on the stage of development.

Going on past the garden there was another building with a pen going out from one side. In this pen three animals were munching on plants that looked like spinach, or as I remember seeing on a farm a field of rape plants. I remember my dad had a field of rape for hogs to graze on one year. Supposed to be very nutritious.

These animals were about the size of a large goat. Actually, they looked a lot like a goat, but they had very short hair all over that reminded me of the the hair on a German Shorthair Pointer dog. Might just have said it was like the milk goats back on Earth. And sure enough that is what they were used for, milk, or something like it.

All other animals I saw apparently had the same kind of genetics: two of everything: two eyes, ears, arms, legs, noses, and of course pairs of legs. If I come across other type animals they will be described in more detail as well as these.

Going back to the house, the flat roof meant that it contained something that turned sunlight into electrical energy. Fred noticed my looking up at it, and explained that the movable items mounted along one side of the roof were mirrors that focused the sun onto the energy cells. The mirrors could change their angle to follow the sun across the sky, and focus the rays on the energy cells.

We had apparently taken up most of the day outside, and got back in to see Wilma fixing a meal. By the time we had eaten the meal, it was starting to get dark already. The short time of this planet's rotation was giving my earth system lots of trouble adjusting.

In time I did get adjusted to the short times of night and day, and the two meals each day. After what I considered twenty-two of this planet's days there was another visit with the authorities. I was notified that a meeting was being scheduled with the council members in three days' time, and to be ready for any questions that would be asked. A car would be sent for me and both Wilma and Fred. This meeting was to determine if freedom would be granted, or if my life would be ended without further attempts to adjust to this new life style.

It turned out that it was a good thing Fred and Wilma went along to that meeting because their testimony convinced the council that I could adjust to both this kind of climate and the quicker change between night and day. The main concern was getting along in society without causing controversy. Fred was very supportive in this part, and even though there had been little interaction with others he convinced them I could adjust nicely. Others had found me a curiosity of course, and my different appearance led to many questions but caused no trouble.

The decision was to let me stay, but with the condition that Fred and Wilma had to see there was more exposure to the culture and increased interaction with others. Then help to find a place where I could be useful. So occasionally we had others come for a visit. Conversation was easy, and there were many questions on what it was like on my home planet, and of course the story of how I got to this planet. Most of them had already heard about it of course but enjoyed hearing it from my point of view. So the story was modified a little to make it more colorful to hear. Soon there were requests to give talks about my planet, and how it was different from this one. I just had to repeat a lot and try to adjust my voice so they could follow what I was trying to say.

With the questions and explanation of how our government worked on Earth, it became obvious I didn't know that much about the government on this planet either. So one evening after the meal Fred was interrogated on it.

This led to some surprises. One especially was very intriguing. There were local councils, then larger district councils, and finally the head continent council. I could understand how that could work of course, but so far I knew nothing about the other continent, or if there was an organization that coordinated laws or regulations between the continents. So I asked that very question.

"I have told you before how the wars between us were very destructive. Many lives were lost, and much property destroyed. So meeting in the middle of the ocean it was agreed that wars and fighting should be eliminated. We have a joint council that meets twice during our solar year to resolve any necessary adjustment" Fred explained.

But once each year there is a contest to pick a king that will become ruler over the whole planet. The council decides all important issues. The King is for ceremonies only, and to make decrees on things like festivals, and holiday functions. He sometimes officiates at weddings and important funerals.

"In deference to our violent past, the king is decided by a person selected by each continent. They meet at a festival where a fight determines the king. Only one person from each continent is involved, so there is no more violence except what takes place in the square on that day. This is a symbolic fight to pick a symbolic King that has no real authority. The continental council is the real authority, and does not interfere with the decisions made on the other continent."

Seeing my still questioning face, Fred explained that a tournament is held on each continent to select their representative to be in the fight. It starts in the local area with the winner going to the district level, and then to the continent level. The continental winner then represents us in the king fight. This is the only physical competition allowed.

"Would you like to go see the start of the next tournament?" Fred asked'

"Yes, I sure would," I replied immediately.

"Okay, I will find out when they start, and where the first ones will be held, and also who is fighting," Fred said. "In the meantime, a trip to the local council will give you an example of how government operates at the lowest basic level."

Now I was getting into the city areas for shopping. As we went along occasional stops were made just to talk with the locals, and learn as much as I could about their everyday life. It was not too different than life on Earth. Surviving as easy, and as comfortable as you liked to be was always in the forefront of any discussions I had with them. I used my different appearance to get a conversation started. When somebody was obviously curious I would simply make a pleasant nod, say hello with a big smile, and ask the location of something. That is all it would take to get it started.

This seemed to be a pathway to getting to be a likable character and the start of a friendship. I was very curious and interested in the females but did not seem to be making any headway in getting to know any. So in desperation, I turned to Fred to help me with any information about them. They didn't seem to be very interested in any males, not even those of their own species... I was really interested to know what they looked like under their clothes, but it didn't seem to be the polite thing to do to inquire about sexual habits and things. So I was pleased that Fred would give me the "father-son" talk at my age. This was the information that Fred told me.

"In the past, the population on our planet increased to the point that it couldn't continue to provide enough food or room. That led to the most severe wars that ravaged our homes. When the two continents decided something had to change, and all war was outlawed, it was determined that population control would be mandatory. Every male baby was immediately sterilized. When they reached the age of 50 solar years and wanted to start a family, then the sterilization would be reversed. This was done only with the males because the operation was more involved in the females, and also more complicated to reverse.

Both the male and female would apply for a permit to have babies once they found the one they wanted to be with. Both would then do genetic testing for any genetic problems. This procedure led to females being very careful in selecting a partner. If the genetics was not favorable, then they could consider somebody else before getting involved with a particular person that they could not keep to share a life together and produce any babies.

"So you see Daniel, you are a different species, so having a baby with you would not even be possible. That does not mean you could never take a partner. You would probably not even have to be sterilized since you are a different species, but to be on the safe side, the council would surely insist you undergo sterilization. You would probably have to be satisfied with a female that was genetically unsuited to have offspring and had already been sterilized.

So when one of our females sees you and takes note of your appearance, you will be ruled out immediately. When they see that thing that you call a nose sticking out in front of your face that will pretty much end any further consideration of you as a mate right there. I'm sure they are impressed with your size, and the rest of your build, but that nose just looks awful. It is so big it is even bigger than both my two noses together." Fred smiled but didn't" laugh at his joke.

"But that is what we breathe with back on Earth. We don't have as much oxygen in the air as you have in the air here, so I need this nose. The ladies will have to accept me as I am, or my destiny is to live with the way I look," was my half-hearted acceptance of his explanation. "But you didn't explain what they looked like when naked."

"And I'm not going to. If anything develops between you and a defective female that is something you will find out for yourself." Fred indicated that was the end of that discussion.

By now I had seen a lot of the residents in this area. They all looked so much alike that it was hard to imagine how a female could be more attractive than another. There was some variation in height, but not nearly as much as you would find in people on Earth. I hadn't paid much attention to the males as they all looked like each other as well. The females all seemed to be out of the same mold. Their looks were so much alike that you would think they were all sisters, and in many cases twins. For the moment I chalked it up to a small gene pool in this area.

The days go by so fast with the speed of this planet's rotation, but time seems to drag on so slowly with the same old daily routine I find myself in now that I've been here for a while. There is little excitement of any kind in my life now, and I wonder if the other choice would have been better. Just choose to be eliminated, and have done with it all. Nobody here would miss me. No chance to leave any progeny for a future. No possibility for any kind of a lasting legacy. I have nothing. I am worth nothing. I can't even pronounce the name of my care takers properly.

"The tournament is taking place in two days," Fred announced. "Would you like to go see it?"

"Sure would," I said. "I've been waiting quite a while to see this."

The day of the fights came around, and Wilma packed a lunch for us as it would last most of the day. The area was busy with spectators when we got there. It wasn't long and we all gathered around the "ring". It really wasn't a ring, but a square about thirty feet on each side. The sides were about three inches high.

The announcer stated the rules of the fight for all to hear as the combatants entered the square. There was no time limit, no periods of rest. The winner would be declared when one of the contestants stepped out of the square, and called an end to it himself, or it would end if he were knocked down, and would not get back to his feet. I don't know how the pair for each fight was chosen.

The fight started, and neither one seemed to be in a big hurry to get going. With no time being kept it looked like they were going to pace themselves looking for a hit that would mean something. Then with no gloves or padding on the hands a lot of hits could gradually break down a hand, and make it too painful to land hard blows.

I wondered how thick their skull bone was, and how much of a blow it would tale to do damage to the brain.

After watching for several minutes it didn't take long to understand that they did not have fast reflexes. I knew they did not have as much muscle or as much strength as Earth people, and by comparison to Earth people they are really slow. So slow it became obvious to me that I could whip either one of these two with one hand tied behind my back.

This fight ended when one of them just sat down, and refused to get up. It was amazing to me that he would quit just like that without an obvious injury, or none that I could see anyway.

The next match was a little more exciting. These two at least got in some blows on each other. There were no knock downs, but when one got hit close to the border of the square, he stumbled over the line, and decided not to get back in bounds. So that bout ended.

Two more matches that ended without so much of a knock down. Just a few hits was all it took to decide. I was puzzled by it, so asked Fred for an explanation. But Fred didn't have an explanation. The entire bracket went the same way until the winner was decided. He would advance to the district competition, and that winner would go to the next bouts at the district level.

I noticed that there were quite a few females standing around the area, and were taking notice of who was winning each match. There were several that were giving me a good-looking over from the backside. There were several comments I could faintly overhear on my good build and solid frame. They would then gradually work their way around to the other opposite side of the square, apparently to get a good look at my front side. Some turned away immediately to move to another side of the square. Others stayed longer and stared at my face for some time before moving off. I knew it was my nose that turned them off. Reminded me of the old saying on Earth that "beauty is only skin deep" but that didn't seem to apply any more here than it did on Earth.

"Oh well, they probably wouldn't be any good in bed anyway," I told myself and promptly dismissed the thought. On the way home, this situation was discussed with Fred.

"What If I didn't have this nose? Would the females have any interest in me as a possible mate? I'm just curious you know. They looked me over pretty good during the fights"

"I saw that," Fred replied. "Thinking you would like to have a mate?"

"Well maybe some time," I stammered feeling a little embarrassed by it. "I can't take it off. Need it to breathe in air. Maybe it could be reduced in size a little. Do you have people and hospitals that can perform operations like that? Wouldn't need to remove much, just enough to make it a little smaller, you know, just enough so it wouldn't be considered repulsive to everybody on the planet."

"We will make inquiries," was all that Fred said, so I concluded that subject was closed for now.

Many solar days went by as we tended to the chores on the farm. As I was getting bored with the daily routine, we discussed my striking out on my own. But in order to be self-supporting, getting a job had to come first. This turned into another big road block. There were many jobs in the area that I could do. The inquiry brought an invitation to interview for the job, but when the interview was finished, nothing ever came of it. So no job was ever in the offering. It didn't take too many times being turned down to realize they would not hire an outsider; especially somebody that was of a different species.

How many times had I seen this done on Earth, but in most cases it didn't affect me, so didn't pay much attention to it. No, if I did notice the discrimination, the realization began to sink in that I did not have the guts to speak up about it. As long as it didn't happen to me there was no concern about it. Now that it is happening to me, is nobody going to speak up for me?

I spoke to Fred about it, but there seemed to be nothing he could do, or even wanted to do. My mind now turned to how to make them pay for this discrimination. Since I did nothing on Earth to make them feel the pain of someone being discriminated against, and now regretted that I had not lifted one finger to help, but now somehow, in spite of my failure to help on Earth, these jerks were going to learn to respect all beings no matter how different.

In the meantime, it was odd jobs, and low-level work was all that was available. Fred and Wilma were very supportive as I continued to live with them. Sometime later Fred announced that I had been turned down on getting my nose rebuilt, as none of the surgeons had any experience with it, and none wanted to be the first to operate, especially on an alien.

Sometime later we learned that our fight champion had lost to the other continent to be the next king. Seemed our continent had not won in nine solar years. Our king house had sat empty all that time. It had been maintained, but nobody went to see it anymore.

"What does the king do," I ask Fred.

"Not much. The king is more of an honor for the continent, rather than a leader that makes rules and regulations. He does things like approve festival days; okay uniforms for different service groups; and approve trade goods as well as coordinate exchange of goods with the other continent," Fred said.

"I am going to compete for the King position next year," I announced to Fred and Wilma. "Getting back in shape will start tomorrow. I used to be in really good shape during the wrestling season. Still didn't make the first team, but I was a good wrestler."

"You may have to get the approval of the council to do that," Fred advised. "Would you like me to present it to the council the next time I'm there?'

"I would prefer to do that myself," I replied.

"That's fine with me. That would be a little embarrassing asking to have an alien represent us at such an important position," Fred opined.

"When they are told that we can have the king on this continent for the first time in nine years, they should jump at the chance. When they see me beat all the competition here, how could they turn me down anyway?" was my reply.

The next day kept me busy in the out buildings looking for things to use for weights in building strength. In between lifting weights, working on chores and care of the farm, then running to round out my conditioning, it all kept me busy.

Fred could see that I went at the work with more enthusiasm, did a little extra at it, and that took some of the work off his hands, and so there was no objection to my efforts. He commented more than once on my speed, and seemed to be impressed with the amount of weight being used for the strength training.

When I asked Fred if he knew the names of any local competitors in that last competition, he didn't know, but offered to find one for me to spar with. It turned out that was not a good idea, for their reflexes were to slow to avoid any hits, and also not strong enough to do any damage when I lowered my defense to take one of their hits just to see how hard it felt. Then when I decided to land a really good hard hit on my opponent he went down like a ton of bricks. At first it looked like he was not going to recover, but thankfully he did gradually recover and staggered to his feet. I apologized several times, but he refused to spar with me again. Word must have made the rounds because nobody would practice with me after that.

Several days later, Fred was going to the big city to meet with the council again. This time to have his crop plan approved. He thought there should be no problems, but this formality had to be done each year. I decided to go along and see what it would take to have an application approved to represent this continent in the competition to be the new king.

When Fred's turn came to go before the Council members, I went in with him. The members recognized me immediately, but instead of talking to me they inquired of Fred how I was doing. Fred gave me a glowing report, and asked that I remain with him and Wilma. The council agreed, and proceeded to consider his farming requirement. That too was agreed on, and we were dismissed to leave, but I quickly spoke up to ask about getting an application for the competition. That was met with a few snickers.

"What can you do to even make it through the competition in this country? The winner of the other continent is usually a better fighter than anyone here. So why would you even want to try?" one of the members asked.

"Because I can win. I've watched some of your matches, and so I know I can win. For the first time in nine years you can have the King here. I can do it for you," I told them.

I wanted to ask them about discrimination and prejudice against myself, and others that were deemed unfit to have a baby, but didn't want to antagonize any of them so thought better of it than to bring it up at this time.

"We can't see our way to having an alien represent us," one of them said. "You may enter the local competition, and if you win we will discuss whether you should compete further."

To leave on a pleasant note, I thanked them for that consideration, and also for letting me stay with Fred and Wilma. Neither Fred nor myself said anything about my workouts to get ready for the local competitions. We left without any other discussion with them on my training, got in the vehicle, and headed home.

I was still a little curious about how these electric cars operated, so had some questions on how the sun power was transformed into electric power. The voltage that was routed to the motors, and how much to the power steering motors, and to operate the brakes. It was a comfortable easy ride, but without warning one of those fungus plants sprang into the air, and the wind blew it right into our path.

I yelled to stop about the same time Fred hit the brakes, but it was too late. The thing hit the front of our car, and spores splattered all over the place. No damage done to the vehicle, but all those spores sure left a mess in the air, and all over the front of the car. Fred started the car going again, and didn't bother to look at any damage. So I had to ask another of my stupid don't understand type of questions.

"Shouldn't we stop to see if there is any damage to the car? How about the fungus? Does it need anything done to it? Fred just smiled as he shook his head.

"The fungus doesn't have any nerves that transmit pain. It was about ready to explode its spores, and you could see that spores just went all over when it was hit, so nothing was left to pick up. They would just blow away with the breeze, and anything left on the car will blow off as we go along," was the explanation.

When we got home I told Wilma about hitting the fungus, and what happened to it, and how it just splattered all over the place.

"Nobody was hurt. I guess not even the fungus was hurt. At least Fred said a fungus didn't have any feelings. Thank God it wasn't any worse," I commented.

"You have mentioned this God several times lately. Tell me what is this

God you keep saying thanks to. I have never heard of it," Wilma asked.

This is when I wished that no comments on God had ever been made for how do I explain what God means on Earth to somebody that this concept is completely foreign to on this planet. But I had to try.

"God is not a real person. God is a belief that many people have. That everything was created by this super power we call God. That it creates our life, and the life of everything else. Even the plants and animals. Even the planet we live on, and everything that lives and grows," was my explanation.

"That must be very handy. When something doesn't go right then you can just blame it on God? What if you plant a seed, and it doesn't grow? Does that mean God wouldn't let it grow? And why would he keep it from growing? Could he do that if he didn't like you?"

Wilma was putting too much thought into this for me to deal with, but I was trying my best to explain it.

"So you can't be born unless God creates you? Why can't you and your partner create the baby yourself? Why does God have to do it for you?" Wilma continued. "What does God put in the female to make a baby grow? What do you need the male for if this God makes a baby grow? How can this God take care of so many females?"

"I don't believe that. Just some people do," I replied, but before I could continue with an explanation Wilma jumped in with her next question.

"If you can't be born without this God, can you die own your own, or do you have to have God do it for you? What if you want to die, but God says no, or what if you don't want to die? Can God make you die anyway?

Such basic questions, but phrased in a way I had never heard before, and I was having a lot of trouble trying to explain in as simple way as possible, yet make it seem understandable and reasonable at the same time. Not able to find the right way to do this, a change in the discussion seemed the best way out of it.

"I have to go find some things to put together for use in my workout to build up my strength," and got up to leave.

"I don't understand this God business. Will you talk with me again on it?" Wilma asked. "It sounds interesting, but it is still confusing to me.

"Yes, we will get into it in more detail some time," I told her, and headed outside to see what I could find to use for weight training. I knew that I had more strength than the local people, but wanted to make sure the muscles were well-conditioned for the next contest. The confusing thoughts of religion were beginning to spin around in my head. I was happy to get away from Wilma's questions but was still having trouble getting away from thinking of how to answer her questions.

It kept nagging at me how getting a job was almost impossible with the way owners and managers were treating me, and I suspected others that didn't measure up to the standards as set by the "Normal" ones were treated the same way. These thoughts at least were thankfully getting me away from religion

To learn more about what would disqualify somebody from getting a job, I went to several different local offices that were managed by local people. Being nice and friendly while just inquiring about the job market, and passing it off as nothing more than learning how this society was different than on Earth soon provided some of the information I really wanted.

As suspected, any one that was found genetically deficient for any reason was listed. An employer could access this record. The genetically approved would always take precedent on getting the job over an "inferior" applicant.

I was able to acquire a few names. Then as time permitted a visit with every one of them confirmed what I already knew. The discrimination was pervasive throughout this whole society. The interesting part of this is that since these "inferiors" couldn't get a regular job, so in order to make a comfortable existence they turned to schooling and training in the research and development area. The research companies were glad to have them because they would put in long hours to develop something they were interested in as their social life outside of work was practically nonexistent. Knowing this I concentrated my effort at developing friendships, and support for them, and for myself. It took a lot of time with my work on the farm, and travel to and from the villages. Since I didn't have a vehicle the trip into a village was made by running. This was the excuse to myself that the run was one of the keys to fitness. Combined with the weights at home this would be all the work needed to be in tip top shape.

It seemed with each trip to visit a fellow "inferior" they knew the name of another person that would be in the same category. One after another as the contacts added up, each and every one felt the same way. It was very unfair. Each one would somehow like to retaliate, but were so isolated, and outnumbered that they just put up with it to cope the best way they could to survive.

The time came for the tournament to start on the path to being selected as the continents representative for the king fight. There was just four of us entered. One of them was the winner from last year that went on to the district and then to the finals where he lost.

As luck would have it I drew his name for the first match. Fred and Wilma both came with me, and of course, they were the only ones cheering for me to win. There were a few sneers, and laughs as I entered the square, and Fred and Wilma got a few stares as well for cheering for me.

The fight got started, and I held back a little to see what my opponent would start to do. As he came in closer and drew back to land a blow, I just fainted with my left hand, and then hit him with a good solid right to the left side of his jaw. This is a blow designed to snap the head to the side, and bounce the brain against the side of the skull. When done with a hard enough blow it usually means a knockout, or some serious effect on whoever takes that kind of blow. But even this firm, but not that hard, of blow, put this opponent down and out. He did not say he was done, and could not even get up to say he was done.

I had not meant to hit him so hard that it rendered him unconscious, but when his senses returned enough to realize what happened he agreed to quit, and I was the winner.

This brought the realization that my strength and reflexes were so much superior that succeeding matches would have to be done with more care on my part. I wondered if their skull and brain padding could also be more delicate. With my natural advantage, it could lead to strong objections on being allowed to compete.

So the match for the local final would have to last longer and appear to give my opponent some possibility of winning. Being curious how much strength he had, and how hard he could hit, several punches were allowed to land, but not at a straight-on blow. There was a little more power there than I had expected, but not enough to be much of a concern.

It would have been as easy to knock him out as the first one, but didn't want to do damage to his head, or body, so just hit him enough times, and avoiding any more of his punches which would lead to the realization there was no chance of his winning. He readily gave up and called it quits. There didn't seem to be any shame or disgrace to recognize you were out matched, and just call it quits. No referee needed to stop a fight. That was left to the combatants to decide for themselves.

Apparently it didn't take much strength to overcome this small planet's gravity to lift anything so muscles were not as well developed. I have no idea why their reflexes are so much slower. Just the genetics I suppose.

It didn't take long for the result of the competition to spread throughout the community. I took this opportunity to call a little "get together" with some of my "inferior" friends. While we were discussing my win at the tournament I took this opportunity to mention that it was more likely that we were the superior ones instead of the others.

This idea was taking hold, and there was some discussion on the way we were superior. For the most part, the jobs held by those of us that were not suitable for taking a partner and procreating were of the more intellectual kind that took more schooling and training. Maybe the general public looked down on us for just that one reason, or maybe there was more jealousy than we thought about our having the better jobs. If only there was a lot more of them. We agreed to meet again in the near future for the common interest and for the fun of laughter at making jokes about our luck in life. This was a good opportunity for each of us to learn about each other's background and our role in this society.

Asking the one I named Harry why the genetic testing was done to everyone in the first place, I asked for an explanation, and got a detailed version of why all of the planet's inhabitants had to go do it.

"Many generations ago many of the new born had problems with disfigurement or health issues. When the problem was severe, the baby usually died shortly after birth. Those that did not die early in their life needed special care. This took a lot of time, and often was so severe it was a drain on resources that led to extreme measures taken by a parent to survive the ordeal. It was not uncommon for the life of the child to be forfeit as a result. Those that did live to maturity wanted to find a partner that would accept them and often it would be somebody with the same or similar defect. When they would procreate the offspring would invariable have the same defect?

When the ability to inspect the genetic makeup of a body was developed, the Council made a law for genetic testing of every body, and a person could only procreate if their genes passed inspection to be certified clear of any defect. This law has led to a great reduction in birth defects.

The unintended result is that it has led to a population that resembles each other so closely that we all look pretty much the same. I'm sure you have noticed that there is not much variation in size, and our skin color lacks variation. So you see that is why we are all a very attractive species," Harry laughed at his last comment as we broke up to leave.

It was about twenty days later that the district competition was to take place. My newly found friends went with me to the meet site. When it was my turn to go into the square, they cheered loudly to support my winning. They stayed for the entire tournament and gave a rousing cheer at my being crowned the tournament champion. This meant I would be going to the finals to fight for the continental championship.

Day by day we picked up more members for our group meetings. The group was getting too large to meet in each other's home, so a building was reserved to hold our meetings. A committee was picked to visit each of the businesses that employed many people for the purpose of encouraging the management to hire an "inferior" if their resume and skills were superior to the other applicants. At the next meeting, the committee would report the result of the visit.

During one of the days during that time period, there was a notice sent to Fred that I should appear before the Council for further consideration of my request. Fred made the appointment, and when the time came we went in his vehicle again. I teased him all the way about not hitting a fungus again and messing up the whole front of his car. This teasing him helped keep my mind off why the Council wanted to see me.

They had decided that if I won the final competition, and was going to represent our country in the king fight, then I should look more like one of them. I could have my nose reduced in size. It would not be as small as one of their noses, but at least it wouldn't be so large as to be grounds for disqualification, and the country would not be held up for ridicule at having such an ugly competitor. But this nose reduction would take place only if I won the final match.

I readily agreed to those conditions, so we were in a very good mood as we left for home to tell Wilma the news.

"Do you think I should tell my friends about the noise reduction Fred?" I asked. "Or should I wait until after the finals, since they won't do the operation anyway unless I win?"

"That's up to you and your judgment. You should know your friends better than me, but if it were me I would make sure of the win before bragging about it," Fred replied.

I had relied on Fred's advice and council many times, so his guidance on this matter would be seriously considered. It didn't take too many days of thinking about it to follow this advice. The friends will have to wait until after the fights before being told. I know they think of my one big nose sticking out the middle of my face is rather revolting when compared to their two tiny noses, but they still look to me as their leader. That is where they get the courage to go to companies to try to convince the management that many of these "inferiors" would be an outstanding employee. All they need to know is that somebody is behind their efforts, and at this time that is my role. For a manager to see me in person with this unacceptable nose would get us nowhere in negotiations with the company.

Finally, after the third visit with one company, the manager agreed to hire one of the "inferiors" to work as a coordinator of materials handling. We all were jubilant at finally reaching this goal, even such a small goal as one person. This was encouraging, so we were determined to continue with our efforts.

Then the day came to head to the Capitol for the final tournament. Fred and Wilma both went with me, and most of my friends also went. The matches were a little more even than the earlier ones, but none of my opponents could come close to matching my ability. I kind of felt a little sorry for them, as it was not their fault that speed, strength, and reflexes was not their natural condition.

I was crowned the continental champion. There was reservation about cheering for this "ugly" foreigner from most of the crowd, but my contingent gave me a long enthusiastic cheer.

As soon as the celebration quieted down, members of the Council wanted a meeting. They confirmed my nose job had been approved, and already scheduled in case I did win the championship. The operation was to be done tomorrow, as time would be needed for the wound to heal. The king fight was only forty days away, and the operation plus any corrective surgery would need that much time to heal enough to withstand any blows.

Fred. Wilma and the friends left for home without me, and the council members took me to a building to stay for the night, and then take me to the surgery room the next day. It was a sleepless night worrying about how all this would turn out.

The next day the first thing I wanted to know was how it was going to be done. The surgeons assured me they had the skills to do the job, and I would be a much better-looking person for it.

"What do you use to deaden the pain? What kind of anesthetic to you use?" I wanted to know.

"We can eliminate any pain during the procedure with a combination of sound and magnet wave lengths. The frequencies will be determined by the density of the area that is being worked on at the time. You will be immobilized before the procedure begins. You will be able to hear us, and ask or answer a question," the surgeon replied. "Now hold still."

I didn't feel much of anything while they were trimming my nose down to size, but a few moments later it hurt like hell for about ten minutes until the stuff they put on it began to numb the pain. A large bandage went over the area. I was given specific instruction not to remove it, and keep your fingers away. Don't even touch the bandage.

"The caretakers here will be in charge now, so you do exactly as they say, and everything will heal up just fine," the one Doc said as I was wheeled out of the operating room and into another one. A message will be sent to your parents that you are doing fine."

I didn't want to correct him on Fred and Wilma not being my mom and dad, as it was not necessary. They were beginning to seem like my mom and dad anyway. This caused thoughts of my real parents to come flooding into my mind. What were they doing? Had they given up looking for me? Since there was no body to be found did they hope I was still alive somewhere? Where had I gone? Was I kidnapped? Did they have the rifle that was left leaning up against the tree? I missed them with all my heart, and assuming they missed me as much started me feeling guilty about leaving them.

The impossibility of returning to Earth put me into a depression. Then my thoughts turned to anger at being in this situation. The snide way these people looked at me; the insults on my one big nose; the discrimination and prejudice when trying to find a paying job.

I will show them. I am strong. I am smart. My fellow "inferiors" and myself, we will show them who is inferior. I will become the King. Then what powers the King has will be used for the benefit of my people.

We have a good start with my group on lobbying for jobs. It looks like this is the time to start employing some tactics I remember from Earth history. The tactics of labor boss John L. Lewis, and Jimmie Hoffa brought big changes in the labor force. So now it's time to get serious on the use of group strength to change the labor force here. I am the leader of my group of friends, and I shall be the leader of the movement that will go down in history on this planet.

The nose healed enough to go home. The caretakers removed the big bandage for the last time. A mirror was brought in so I could see my new smaller nose for the first time before covering it with a smaller bandage. It looked ridiculous to me, but it did look more like the ones on these people. With only one nose in the middle of the face instead of the two matching ones the natives have, it still stood out like a headlight that called for attention. With instructions on how to care for it until completely healed, the surgical unit sent me home in one of their vehicles. On the way, the driver explained that since I might be a future very important figure is why such a courtesy of a private ride was being extended to me.

Fred and Wilma seemed very glad to see me as they looked closely at the bandage over my nose, so I gave them a brief account of how it went, and the effort that was taken there to make me comfortable. Wilma looked through the things to use on the nose until it was completely healed. No doubt she was going to take good care to make sure the bandage was changed as often as needed. Her motherly instincts made me realize that she would make a good caretaker of anyone in her home. Fred was equally concerned about the healing process, and every day asked if the bandage needed changing. He had even been out in the shed putting together some combination of weights for me to use. I thanked him for doing that, but he shrugged it off as just wanting to make sure I was in good physical condition for the king fight that was coming up in a few more days. The last bandage was removed, and they both gave a good inspection of the new nose.

In Fred's opinion, it looked pretty solid and should hold up okay since it still had some time to firm up before the matches started. I had been working the weights and running, so conditioning had not suffered much while the nose healed.

I think Wilma had been looking forward to the day when I would be more presentable to the public, for the next day we were informed that there would be some visitors coming tonight to test out a new recipe she was going to use. That was very much okay with me as that would show whether my new face would be reasonably acceptable to the locals.

It was a bit of a surprise to me when the people that came the next day was a man and wife along with their reasonable attractive daughter. So I knew right away that Wilma was not only wanting to help me fit in this society, but would like to see their neighbor's daughter and myself possibly be a match to start our own home. What I wasn't sure of was this just to get me out of the house, or to really start our own home. She probably hadn't considered it would be impossible to have children of our own as the genetics would be impossible. So I couldn't help but wonder if this lady had been given the physical, and whatever they check, and found to be genetically defective.

I didn't much care at this point for my mind was on getting another meeting before the fight to plan strategy on convincing more managers to see things our way. Besides, what female on this planet would have anything to do with an alien? When I saw Wilma staring at me with that look brought me back to the realization that this little meeting was for my benefit. So be more pleasant, and talk nicely with these people.

I wished it would be easier to pronounce the names of these people, but there were sounds and variations in the voices that I could not make, so to keep straight who was who, the names I used was of the friends and relatives on Earth. So this female became Cindy. They did much better at pronouncing my name of Daniel than I did with their names.

"Cindy, have you been through all of your schoolings yet?" I asked.

"Yes, I have. My certification is in general mechanics of laser machines," she replied.

"Have you taken all of your studies?" was my next question as my head turned to wondering if she could be of any help with this damn discrimination of the defective class.

"Yes, I think I have. The studies in that field are no longer available to me," she said which was true, but not for the reasons she might expect.

"I know you are an alien. Wilma has told me all about it. Also that you had only one air intake instead of two like we have. That you have just had the one big intake reduced in size to look more like us," Cindy commented. "What else should I know about you?"

"Did she tell you about the fight to become the King? Did she tell you that I have already won the continental championship and will be representing this continent in the fight to be the King?" I asked.

"Yes she did. I hope you win, and become our King. We have not had the King here for many years. It would be a big celebration when you came back with the badge. The other continent seems to have all the winners. I think the reason is because they train for it all year long. Do you train for it?" Cindy asked.

"Yes, I do work out with weight, and run for conditioning. Fred put together some additional weight while I was having the operation. Would you like to see what the things are that is used for weight?" I asked.

"Yes if you don't mind. We don't do a lot of working with weights. It is considered the work of somebody with low intelligence. Oh I'm sorry. I didn't mean you have low intelligence. I know you have a goal, and that is not considered low intelligence," Cindy said as she continued with some giggles.

"That's okay, I sometimes wonder if I'm not a bit on the short side with brains to work with weights," I replied as we went out the door.

Wilma and the neighbor lady took note of us as we went out, and I'm sure they shared a smile at seeing us go together, and wonder just what that was all about. It seems there are some similar activities on Earth as well, but not for the same intention. This was informational and not romantic.

Showing Cindy the weights being used, and how they were used to develop certain muscles she wanted to try it too. Using lighter weights she did pretty good for a beginner. It was the first time since my arrival on this planet that I had fun and laughter with a female, and with all things, a bunch of junk used for weights.

We were still laughing as we went back into the house, which got the attention of the ladies of course. Fred and the other male were sitting at a table discussing farming I supposed when Fred turned to ask.

"What is so funny with you two? Did you get lost out there?"

Cindy was still grinning as she told me about trying to lift some of those contraptions you call weights.

"You could smash a foot or something if one of those things dropped on you. I wouldn't want to do that for a living," she added.

As the four of them went back to their own discussions, Cindy and I sat across from each other at the table. I took that opportunity to look closely at her hands that lay on the table in front of her. I reached out with my hand and took hold of one of hers. Gently lifting it off the table and turning it over from one side to the other gave me the opportunity to study those two thumbs, one on either side of the four fingers. I found the grasping ability of that combination of fingers and thumbs amazingly dexterous and strong.

"I hope you don't mind if I look at your hands. It is amazing to me how you can use them in handling things. Here, look at mine. Four fingers and only one thumb. A distinct disadvantage compared to your hand," I commented.

I knew it was being a bit forward, but in addition to marveling at how her hand was built, I was enjoying holding it for a little longer than necessary. She noticed I was deliberately taking time, and put her hand back down on the table. Feeling a little embarrassed that she had noticed, my hand quickly went back down to the table top as well.

"Tell me about the females on your planet. How do they differ from us, other than only one thumb?" Cindy asked.

"And only one air intake," I laughed at my own comment. "If you had seen my nose before the operation to reduce its size, then you can imagine how the females looked with the larger nose. On planet Earth that is the natural look, and many of their noses are quite attractive. Some are larger, some are smaller. Some are wider, others much narrower. Some are narrow and pointed, while others are broad and bulky. So you see there is such a variety that it depends on the rest of the face which nose looks good with it.

And the color. There is a great variety of skin color. All the way from black to brown to white, with all shades in between. Same with the color of their hair. As you can see, mine is kind of brown. Some even have red hair, and of course, here to there is a lot of variation in color. Some have let their hair grow long, while others have cut it to a shorter length.

On average, the females are taller than the ones here. They have more of an hour glass shape as I showed with my hands what that would be. There is a big difference in size as well. Some are much smaller than me, while others are larger than me. The shape of a body can vary by a lot too, but the shape I like to see is similar to what we call an hour glass. Then of course I had to explain what an hour glass was and how it worked as a time keeper.

Seeing her puzzled look, the word hour glass didn't register, so I traced out the shape with my hands to show that kind of shape again. She really wasn't getting the picture of a girl's form, so quickly changed the topic.

"You have a very interesting education in laser technology. What company are you working for?" I asked.

"I don't have a job. Once in a while, a research position comes up when a company needs a minor problem checked out. I thought you were informed that the genetic laboratory had labeled me as an "inferior". I have a genetic defect, so have been sterilized," Cindy explained. "So a company will only give me a temporary job when extra help is needed, or a problem needs to be addressed."

"Oh, no I was not told. I am so sorry."

"Don't feel sorry for me. At first, it was very disappointing that there would be no offspring in my future, but then realizing how much more freedom would be available to go places; to continue with education in the field of my choice; to associate with any one I choose; avoid the responsibility of bearing the burden of having, and then raising offspring. No, don't feel sorry for me. I rejoice at the liberation," Cindy smiled at the final comment of liberation.

My thoughts on what she had said started me thinking. Maybe she would like to join our group at the next meeting. So decided to briefly explain.

"Some of us that are all labeled "inferiors" get together on a regular basis to enjoy some fun and laughter, swap stories, and have a couple of drinks. Then we get down to the business of the best way to convince company managers to hire everybody based solely on their ability to perform the task for which they were applying. We try to make them understand that discrimination of "inferiors" is not right. So far we have convinced one manager. He has hired one of us. I know it's not much, but we have just gotten started. Will you go to the next meeting with me?" I asked.

"I don't know yet, but if it is possible, we will let you know. Thank you for asking. Sounds like a very noble effort," she replied.

Looking up at that point, I noticed the others were paying attention to our discussion.

Her dad got up and announced it was time for them to leave, so the rest of us followed suite. I said the courteous things then, commenting on it was nice to meet them, and maybe we could do this again sometime. That was the first time to use the social skills Wilma had taught me. We went to the door with them, but did not go out of the house. Wilma had said a guest would invite you out if they turned to look at you after they were out the door. I had no idea why but if Wilma said that is the way it works then okay, no problem for me.

I learned later that to go outside with guests that were leaving meant that you wanted them to go, and this would give the impression that you were almost trying to push them into their car to get them off of your place. That didn't make much sense to me, but there was a lot of stuff that didn't make sense. I guess that is no different on Earth where certain actions signified some meaningless little thing that an alien wouldn't understand.

The very next day Cindy sent word that she would like to go to the next meeting, and please let her know when and where. I called her on the "talkie thing" (which was the telephone to me, but a word I couldn't reproduce the right sound, so it will always be referred to as telephone) with the information I write on it.

"I do not have a vehicle and have to walk there. I'm sorry that I can't take you, but it would be a pleasure to have you walk with me," I said.

"Then you can ride with me. I will come by your place in plenty of time to get there before the meeting starts," she replied.

And she did. Her vehicle was also a solar powered car as everything seemed to be powered by the sun. When the sun was not shinning, the power was from an electric storage unit that continued to build up power until it reached capacity. These were the quietest vehicles I had ever heard. Even the purr of the electric engines were very quiet. Couldn't even hear it when the wheels on the road were noisy.

There were several members already there when we pulled up to park in front of the building. Getting out of the car it was immediately noticed that my old nose was gone.

"Hey, come take a look at this," one of them yelled. "He actually did it. He said it was going to get done some time ago, and darned if he didn't do it. It still looks funny, and he still has only one, but it's much better than before. You could almost pass for one of us if you got rid of the hair he teased me.

So they all came running to see what my nose looked like now. As they stood around gawking at my face, Cindy got out of the car. Then one of them noticed her, and the object of their attention changed immediately.

I started to introduce Cindy to all my friends, but the struggle to make the right sounds brought a big laugh from most of them, and some teasing about how I butchered the pronunciation. So "George" offered to do it. With their names properly pronounced, Cindy did the correction on her name. I never did know George well enough to learn his name or to manage to pronounce it correctly.

We all went inside to wait for the others to show up. I got the meeting started, if anybody had news they wanted to share, it was done first. There had been some applications for employment, but nobody had gotten a job. Our discussion turned to what could be done to improve our situation. Other than to keep talking, and recruiting more members there was not much we could do at this time. It was suggested that if we were going to recruit more members then we should have a structure to our organization; elect a leader and several secondary leaders.

Since I had been the instigator, and led most of the earlier discussions, selecting me would be an obvious choice. But I looked too different, so made it clear at the start that I should not be the leader. I could accept a position as a minor leader, and still, be available for discussions as well as planning sessions.

There was one more announcement I needed to make before we ended. So I told them that in six days the fight to be King would take place. I would be representing our continent in the fight. If I was the winner, then as King anything that could be done for the "inferiors" would be done. At this time I have not been told what the responsibilities of a King would be, but it will be carried out to the best of my ability. There was a loud cheer, and best wishes offered to do my best.

Cindy had sat there the whole time without getting in on the conversation. On the way back home she offered to do what she could to recruit some of the females that she knew who were also "inferiors". She remained quiet for an unusually long period of time, and I could tell there was something she was very concerned about. Finally breaking her silence she spoke up.

"Has anyone told you about what to expect from the ones on the other continent? Have you been told how and why they win most of the contests? You don't even know what they look like do you? Has anyone told you what the inhabitants on the other continent are like?"

"No! Wait a minute. What the heck should I know? Aren't they just like everybody here? I get the feeling there is something I better know before going there for the contest. So give me the information now as I sure don't want to get into something that I know nothing about," I demanded.

"Someone from the central office should have explained the differences, and what to expect in the competition," she said and hesitated before going on. "They are not built the same way as we are on this continent. When they were in the process of evolving, their ancestors had four legs, and no arms. Over many generations, the front legs moved forward at the same time that arms and hands were developing. As a result, they now have arms at the shoulder position, and also arms and hands at the waist level."

I tried to imagine how this would give them an advantage, or how this extra set of arms would be used. Looking at the expression on my face, she could see that I was expecting further clarification.

"They established some rules on it shortly after the first contest. They could only hit with the two upper arms. The problem for our contestant was there guy would grab onto you with the lower arms, and then hit with the upper arms. You couldn't get away, so you would keep getting hit with the upper hands," she explained. "So do not get in close so you get caught by the lower arms. With your superior height, and longer arms you should stay back further and rely on longer range hits."

"Okay," I said. "Is there anything else to watch out for over there?"

"Well, you should expect to see a lot more discrimination. They make fun of us, and consider themselves much more superior to us on this continent. They are superior to us physically with their extra set of arms and hands, but we are the ones that have made the biggest advancements in the academical world. We are far superior in science, technology, energy, and anything else that requires intelligence." Cindy finished her lecture on it as we had arrived home.

Dropping me off at the house she wished me to do well, do my best, and be a brave competitor.

The six days went by, and with my summons in hand when the car came for me, Fred and Wilma wished me to do well also. I put my arms around Fred for a big hug, and then around Wilma where she also got a kiss on the cheek as I explained. "This is how we do it on Earth

The driver and my guard for the trip looked at this goodbye with amusement. This kind of goodbye was totally foreign to them.

The greeting of strangers on this planet was the same as a goodbye between close relatives and very close friends. You both reached out at arm's length to take each other's arms just below the shoulders in your hands with a firm but comfortable grip as the head was nodded slightly. But I had grown very fond of Fred and Wilma so for me I wanted to show my feeling for them. I knew that was a minor breach of etiquette but was sure they would understand.

The drive on this road had been taken several times so it was familiar. The scenery along the way had been seen several times, but with other things on my mind, it really hadn't been "seen" by me. With so many questions about what was coming up, I didn't even know what to ask, so to take my mind off it, the scenery along the way was deliberately focused on to help keep the nerves calm.

The driver and guard were not very informative either, as when a question was directed at them, all I got was a shrug and a "don't know" or it was wait until we get there, and they will tell you all about it. So no choice but to keep looking at the passing properties. I secretly hoped that one of those fungus things would pop up and get blown right in front of us, to splatter all over the front of the car, and on those two as well. Maybe that would get them to talk. But it didn't happen, so pulled into the port without incident.

There was a contingent of seven important-looking, fancy dressed, guys waiting there to greet us. With the customary grip on the arms and a nod of the head they each introduced themselves. Several containers were loaded on the boat, and we followed it aboard. No ceremony of any kind took place; just get on the boat and go.

The boat had the usual overhead power grid to catch the sun's rays. The pair of mirrors on each end of the boat would self-adjust to follow the sun and focus the rays down on the power grid. The pair at the wrong end of the boat that couldn't focus the rays at the correct angle simply folded down so they would not block the rays. This was not a speed boat, but it moved right along. I would guess about twenty to twenty-five knots.

With the boat underway, all seven of the guys, and myself sat around a table where they got down to the business of the trip. I was informed of how to behave once we got there. The spokesman gave very specific orders for me to follow.

"Do not talk to anybody except us. You cannot pronounce words and names properly, so don't say anything. We do not want to be disgraced or appear to be an ignorant, uneducated continent.

Keep your hat on, and pulled low over your face. Try to keep your nose from being seen. That one air intake in the middle of your face looks terrible. If we could have given you the two smaller ones, the surgeons would have done so, but with your system that was impossible, so keep it out of sight.

Do not react to any comment that you hear about how you look, or any pointing at your face, or any laughter at your appearance.

Do not let your hands be seen. Keep them hidden under your tunic. A hand with only one thumb would be considered a deformity, and people would recoil in discussed if it touched them. There will be light weight covers over the hands for the fight to keep fingers from being poked in an eye, so your hands will be covered at that time.

You know the rules for the fight. It is the same on both continents. At the beginning of this fight though, each contestant one at a time will come to the middle to be introduced. The King will be introduced first. He will stand at attention until the announcer has finished speaking. Then he will turn to march back to his corner. Just watch what he does, and you do the same. When it is time to start, the king comes to the middle first, then you join him, give the standard greeting of respect, turn your back to him, march back to your corner, turn around, and come out fighting. Don't hit him if he is down. When he gets up, then you can hit again. The fight continues as you know from past fights, until one of you goes out of the square, and does not return, or until one of you says they have had enough.

We expect you to win. If you do, their committee will come into the square with the King badge, put the loop around your neck, show the respect greeting, and you will be the new King. We will guide you from that point on.

If you lose, just come back to our corner, and stay there while we do the respects, and we will take it from there."

It seemed to me these guys are really worried about looks, actions, how words are used, and any thing else that would make them look like back country, illiterate bumpkins. But I'll go along with it, and try not to be noticed. I wondered why they hadn't told me anything about the capabilities of these four-handed people. Did they actually want me to lose? Maybe there was embarrassment and shame an alien was representing them and not one of their own.

When the boat docked we were met by a contingent of dignitaries backed by a dozen or so people waving some banners. I was reminded to keep my nose hidden as much as possible. Nobody objected or made an effort to get up close, so we got to our location without creating a scene. But I couldn't help wondering again why nobody in this group escorting me did not mention what I would be up against. Not one word on their extra set of hands and arms. I suspected they actually wanted me to lose. Probably couldn't stand the thought of an alien winning, and replacing one of their own as the king. But I didn't care what they thought. If they don't like me that is their problem, because I don't like them either.

I didn't sleep very well that night, but couldn't complain because the bed was comfortable, and the food was okay. I had gotten used to eating strange stuff by now so didn't even ask what it was: just hoped it didn't poison me. Even that wasn't much of a concern as I was thinking that any means of getting off of this planet would be a relief.

My group of seven was down to just two that next morning. I wondered where they went but didn't ask, since even if I knew, it wouldn't mean much anyway, as this place was so foreign to me that what was located where or how to get to it did not register in my mind.

Soon two of them came with food for the three of us still there. The others had gone out to eat. Of course, the others in my group didn't take me with them as I was not to be seen by the locals. I had to admit to myself that seeing those two locals come in with containers held by four arms, and four hands were all I could to keep from staring at them. They stared back with eyes that concentrated on my big head of hair, and the one nose in the middle of my face. I tried to look unconcerned while the food was laid out for us.

As they left the room, even before the door was completely shut the chatter between them began, with their sounds of laughter following them down the hallway.

Damned discrimination. What's the difference if I am a bit different? This difference does not do them any harm. My needs aren't so different from there's. Damned uppity committee. Why should my looks be such a disgrace to my continent or the ones that came with me? I'm beginning to wish they had never operated on my nose. Should have left it the way it was. That's their problem if they don't like the looks of something. They had done nothing but make me feel inferior and ugly. Ugly is in the eye of the beholder, and if they see ugly that is because they have an ugly mind. This ends today. They can see me however they want, but I am who I am, and that's all that I am, so that is what they will see.

Came time to go to the fight site. My attire consisted of long smooth-type sweatpants and a tight-fitting t-shirt. A robe with a hood went on over that. One of the seven pulled the hood up over my head as we went out the door. About two steps further I brushed it back over my head and down the back of the neck. The guy reached up to pull it back over my head, but I stopped him in mid-effort. As he tried again I gave him a look that surprised him so much that he dropped his hands to his side. Not being sure what to do about it, with a voice as cold as ice he was told to "leave it alone." I didn't care that they had no hair on top of their head, but by God, they were going to see what a nice head of hair looked like today as that hood was not going to cover it up.

We got some stares of surprise, and questioning looks going through the street in a slow-moving vehicle, but I left that hood down. When we got to the arena, the head guy got in front of me and asked to please wear the hood at least until the fight started. So to placate him, and not be too obstinate, it was pulled back up over my head.

If the truth be known, I was staring at them with the same curiosity that they had in me. That extra set of arms from just above the hips really looked weird. It was an unbelievable site seeing some of them walking along holding something in all four hands.

Getting to the square, and looking around the arena, it looked like about ten to fifteen thousand spectators in the stands. It was much smaller than our football stadiums on Earth, but there were not nearly as many inhabitants on this tiny planet as there are on Earth. Of course, almost all of these spectators are from this continent, so the cheers will definitely be for my opponent.

Stepping into the square, the change in the crowd was instantaneous. They rose from their seats in unison, and the lower set of hands began to clap in unison. This seemed to be accompanied by sneers and sounds of derision, but initially, I wasn't sure.

I AM THE ALIEN

"Just ignore it," my main aide said. "They always do that to us. Just want to show how superior they are with their four arms."

The preliminary format went exactly as was described on the boat. Of course, every time the champ from last year went to the center a cheer went up from the crowd, which was expected. A smattering of polite cheers each time for me was also as expected.

There were some gasps of surprise and shock as I threw the hood back off my head. Then the sneers and what I surmised as derogatory comments followed.

Going back to our corner the last time, the lightweight gloves were put over my hands. I'm not sure what they were made of, but it seemed similar to those light cloth gloves we use on Earth to keep our hands free of dirt. There was no padding of any kind in them. Since these gloves were made for the people here, they had two thumbs. The missing thumb part hung loose, so would just flop around a little when my hand moved.

Time to get it on, so flipping the hood back over my shoulders, and shaking it down off my arms I moved to the center to face him. A gasp went up from the crowd again as they could now see my body with only one set of arms. They could now see my face, and nose, and hair. A rush of talk and pointing went through the stands. This was followed by laughter, then insults. I couldn't make out a lot of what was called out at me, but enough was understood to make me angry. The discrimination got to me all over again. I would show them. You may think I'm a freak with one nose, and one thumb, but this freak can whip your ass any day of the week. The match went about the way I expected; just like the ones on the other continent. Their opponent was slower of reflexes and had even a little less strength than had been said by my group.

The insulting calls continued as I turned, walked back to my corner, pulled off the gloves, and demanded that my corner guy turn a thumb inside out, and push it back into the glove. Since the match wouldn't end until I said so, or went out of the square to quit, I held up a gloveless hand to motion the opponent to stay back and wait. When the thumbs were out of sight I turned to face him again. The insulting calls really picked up in volume as they saw my one thumb hands.

Okay I thought, now it's my turn to give it to you. Let's see if you can take it as good as you give it. Back home in the preliminary matches the opponents were given the chance to see they were outgunned, and could quit without getting hurt. But these bastards here were going to pay.

Before he could get off one punch I sent a solid shot across his left chin. He was out before he hit the ground.

A gasp went up from the crowd, then went totally silent as he fell to the ground. Backing up a couple of steps and waiting for him to get up, a bit of compassion went over me. Maybe he will wake up and say enough so he won't get hit again. It's not his fault this crowd acts the way they do.

After a few moments his senses returned, and he got up. I waited a few moments and told him to quit before he really gets hurt. But the crowd wasn't having it. They started yelling for him to get after me. Throw some good ones. He started for me again trying to grab onto me with his two lower hands. Quickly taking a step back put him just out of reach to get ahold of anything. Not knowing just how to handle this, it now seemed the thing to do was put him down again hard enough he wouldn't want to get up. He got in close, and tried to grab onto me again with his lower arms, but all that was needed to keep from any contact was to back up a step. The crowd kept yelling, and he stopped to yell out some insults at me. It was enough of this treatment, so decided to end it before the insults got worse. So I did. Another shot across the left side again was all it took. By the time he got up again, the brain was so fuzzy he knew there was no way he could continue. So the fight was over. I was the King. The ceremony was short and apparently to the point. The badge of the king was put around my neck. There was no applause or cheers of any kind; just a bunch of shocked unbelieving spectators. For me, it was an unbelievable feeling. King! Me a King! Let them laugh now. I was their King.

The place was silent as we left. Going back to our room we packed our things and headed for the boat. It was not going to be a good feeling staying here in the middle of people who had just lost their King. That put a stop to their snide comments. I reveled in the feeling that the loss of their King could put such a damper on their mood.

While on the boat, the word was sent back to our continent of the outcome. For the first time in many years our continent had a King. Now thoughts turned to what happen next. So with all of us gathered around the table replenishing ourselves with food and drink we rejoiced at the win.

"Okay, what are my duties? What am I expected to do?" I asked.

"You don't need to do much of anything. Being the King is basically a figurehead position. The Council will determine the laws and raise the money to maintain the country. You can determine almost everything else that falls outside their scope of jurisdiction. You are to get the approval of the Council before issuing any edicts. Most of the things presented to you would be to set dates for certain celebrations; awards and honors for accomplishments; signing various certificates, and things like that," they explained.

"There will be a big welcome for you at the dock. They already know of course, and information travels fast, so be ready for a big mob to congratulate you," one of the others added. Get some things ready to say as you will be expected to make a speech right after the Council head authority finishes with his. Then a welcome committee will escort you to the King house, and answer any questions you would have on how it all works."

"You mean I get a house of my own to live in?"

"Yes, but only for as long as you are the King. When you are defeated at one of the contests then the house is no longer yours, and you must leave. You also will have a housekeeper, a cook, a groundskeeper, and a chauffeur. If you need anything else, send the request to the Council."

As our boat pulled up to the dock the cheering had already started and continued to grow louder as we got off the boat. A pathway cleared through the crowd to a platform. Several people were waiting there to greet us.

Speeches were given, and cheers rang out after each one. I didn't understand much of what was said for the language came too fast for me to follow it all, and the crowd noise didn't help in hearing the complete sound of their enunciation. I wouldn't have known when it was my turn, but they looked at me with expectation, to I stepped to the microphone and uttered something about how proud I was to represent such a fine country, and it was such a privilege to win back the King status for such nice people.

With one on each side taking my hand, and two more followings, this committee escorted me to the house. I noticed that occasionally the two hand holders had to glance down at my "one thumb "hand. But neither one flinched at holding it or showed any sign of revulsion. Maybe being the King was an influence on being accepted as one of them. It did much to improve my outlook on being with these people.

Standing at attention just in front of the entrance to the house were the four servants that were mentioned on the boat. One by one they stepped forward to introduce themselves. I did the standard two-handed greeting with each of them.

Knowing how it felt to be looked down on, my greeting was careful to indicate proper respect. As each greeting was done they said their name. I understood most of it, but can't make out some of the sounds, so again just resorted to using names I was familiar with on Earth. So the housekeeper became Jane. The cook became a cookie. The groundskeeper was renamed Blackie, and the car guy became Speedy.

At first, people were a little upset that I couldn't properly pronounce their name, but when they understood that my vocal cords were not the same as theirs which made it impossible for me to reproduce the exact sounds, then my handicap was more acceptable. In time my four house helpers could feel free to make a little fun about it, but also enjoyed teasing each other about the Earth names I had given them, and used it when I was in earshot of their talking to each other.

As soon as the four had taken me on a tour of the house, and their duties explained, I got on the phone to call Fred and Wilma. They already knew of course and congratulated me on the win. They promised to come for a visit the next time a trip to the capitol was needed.

It soon became apparent that it is good to be the king saying on Earth is certainly true. Food was always available. Clean clothes appeared in the closet like magic. The grounds were beautiful; groomed and trimmed to perfection. The vehicle was always ready to go, and where ever I wanted to go. Speedy insisted he would always be ready.

Two of the Council members came for a visit the second day to see how I was doing, and if anything should be changed or added.

"I am fine, but there is one thing I need to know. Who is paying for all this? Where is the money coming from to pay the staff here?" I asked.

"Don't worry about that," one of them replied. The money is set aside every year by the Council to maintain and run this house, and provide for whatever the King wants to do. So you see that after nine years there is a significant amount of funds set aside to do everything that's needed. If you go over budget on anything we will let you know. In the meantime do not worry about it."

Knowing that was a relief of course. Seemed like there was no responsibility on my part to cover any of the expenses. It was also a relief that the business of running the country would not depend on me. At this point, I knew nothing about how the laws were made, or even what laws were already in force. In time in a course, I will know how this government operates, and where the power really is located. The Council is considered the governing body, but there are forces working to shape the Council's decisions.

One of my first executive orders will be to make discrimination in hiring by companies illegal. To do this properly there must be a penalty of some kind for those companies that break the law. I've been told a new regulation must go through the Council. If that is the case, then it will take smaller steps first before trying to get a law established.

NOTE TO WHOEVER FINDS THIS TREATISE AND READS IT.

My attempt here is to give you a picture of what is happening, and my thinking about the how and why of my efforts to adjust in order to be successful in a totally new environment. Sometimes I can give you a blow-by-blow account as it happens. When there is no time to give you immediate action, my writing will have to be as I remember it. I hope this is recovered by Earth explorers as they would probably be be the only ones that could read it.}

I called my "inferior group" leader today to ask that they come for a visit, so we could discuss the procedure of how to advance our plans to decrease discrimination. They wasted no time getting here. That next morning their car showed up right after breakfast was eaten, and Cookie was still clearing the table. They got there so soon, I was sure they had not taken the time to have breakfast themselves. Turning to Cookie as she removed the last of the breakfast dishes I reluctantly asked her if it would be possible to serve my visitors something to eat. She indicated yes, and hurried off to the kitchen while I went to the door just as Jane got there too. She barged in ahead of me as she informed me that answering the door was part of her job indicating with a wave of her hand that I should get out of the way. Dutifully stepping back out of the way, she opened the door to let them in.

It was a joyful reunion for the five of us. With greetings over, my intentions to introduce Jane disappeared with her as she had already left. We all sat down around the table where so many questions came at me all at once I had trouble keeping up with them.

Before I had gotten through telling about the fight on the other continent, Cookie came in with a tray of snacks which was obviously welcomed by my friends for it didn't take long to see an empty tray sitting there. The account of the discrimination and negativity that any of us would have to endure if we lived there was worse than what I had seen here. So something should be done to counteract it on both sides.

"What would you suggest be done to make improvements," Sam asked. "What can we do to get started?"

"Okay, my first official act here will be to make discrimination illegal. Not just against "inferiors", but against anybody that is different," I announced.

"Whoa, better think that through a bit first," Paul declared. "That is going to be a little too blunt for the Council to digest. If you would make a public announcement first, and then go to a few cities to present your case that would be more palatable for people to accept, or at least start them thinking about it. Then with that kind of publicity, it would be a lot easier for us to get in to talk with managers."

"Wow, thanks, I hadn't thought of it quite that way. Maybe you are right. I should get out to travel the rest of the country anyway. See what it's like. Give them time to know me, and find out I'm just trying to help everybody get along and find acceptance in their community," I replied.

The others agreed with Paul, so that would be our approach to it. Everybody seemed pleased that finally, a plan of action was in the works, and we spent the next several hours on how each of us would put our personal efforts into spreading the word into new areas. We would not appeal to just those labeled "inferior" but to everybody that could have some sympathy for the way others were treated.

I asked them to select somebody from our local members that would volunteer to advise and accompany me on where we should go next. It would help tremendously if whoever they picked could also make arrangements for a place to give speeches. I would definitely need help on the locations and the language. If we could get some fellow supporters to be in the audience, and cheer at the right places I'm sure others would join in as a positive response seems to be contagious in a group setting.

The day was winding down so rather than have them driving home in the dark, I called both Cookie and Jane to to see if it would be alright to have my friends stay overnight, so they could drive home in daylight.

"That would be okay with me," Jane said. "I don't have enough beds so two of you will have to share."

"What would everybody want to eat this evening?" Cookie asked. "You gentlemen decide and that is what I will fix. I'm doing something special for each of you."

They all looked at me to settle what we would have. I was beginning to grasp the power of being a King. The title alone brought respect and deference.

"Whatever you think would be suitable, and not a lot of work would be okay," I told Cookie. "Nothing special now, just something easy for you."

I wondered what her reaction would be if she had been given instructions on how to fix something considered a delicacy that would be a lot of work for her, but discretion and respect for the job won out over curiosity so I let it go with that simple and easy comment.

So the talk went on into the night hashing over the possibilities. Cookie did fix a nice meal for us for which we all gave thanks. Jane had everything ready for the evening. We each in turn went through the sanitizer, put on the bedclothes that Jane had laid out on the beds, and we all had a pretty good night's sleep.

A light breakfast the next morning, the group headed back home, and I moved into the office to organize my thoughts on presenting my case to the Council. There was no intention of divulging what I really wanted to speak on. They would be told this was a get-acquainted tour of the country, and a bit of a pep talk on building on an already good society.

I figured Speedy would have a map since he was the driver. Some time was spent studying where the larger cities were located, and the roads leading into and out of them. I needed some help determining what the various symbols meant on the map, so-called on Speedy to explain what the ones I didn't understand were about. While talking with Speedy on the maps, I figured some questions on driving were in order.

"Speedy, what does it take to drive a vehicle? Do you need a license, or a permit of some kind?" I asked him.

"Yes, you sure do. There is a driving course that contains all the things you need to know. If you can drive a car through the entire course without making any mistakes, you will be awarded a permit. If you make a mistake driving through it, you would be required to go through it again with an agent to explain why you failed the part where you had the error, and to answer any questions. Then in the next several days you would drive the course again. Then if you make it with no errors, your permit would be issued. A young person would be taught the rules of the road, and do practice driving on the local learning area before taking the test.""

"You don't have to take a written test first to show you know all the rules?"

I asked.

"No. All the rules are manifested in the driving course. You would see all those situations come up right on the course," Speedy replied. "But don't worry about that. I do all the driving for as long as you are here."

"Yes, I suppose that is the case, but I want to get the permit so I can drive myself," I said. "When I'm no longer the King, then who is going to drive me? So I want to be able to drive on my own."

"That's okay then, but while you are the King then I must drive you. Otherwise, I would lose this job," Speedy insisted.

"Okay, okay! The job is yours for as long as I'm here. I won't mess it up for you. However, when the time comes you will help me with the rules and regulations, and using the practice course won't you? I grinned at that thought of a grown man and a King at that, needing help with learning to drive. "The Council will be meeting with me soon, so I'll need a ride there."

"Yes Sir, just let me know when, and I will be ready," Speedy said, and that ended that conversation for the time being.

I still can't pronounce the name right they call my position here, so I'll just keep using the word king since it's easy to write, and easy for me to fancy being a King.

 The phone call was made to the Council for a meeting. The secretary took my request and told me to wait for their call. Three days later the secretary did call and set a time for me to be there.

"Hey Speedy, fire up the car. Time to go to the Council for my meeting," I yelled out the door.

"Keep your pants on. I've got it started. Be there right away," he yelled back. It seems like Speedy is enjoying using some Earth slang he's caught me saying occasionally.

This was getting to be kind of fun, being able to go back and force kidding each other. And some of the friendly insults were similar to those on

Earth. No stuffed shirt attitude here. I think my relaxed attitude, and not being demanding was having the same effect on these servants. They were never referred to as servants even though that is what they were. Calling them that seemed to me to be about the same as saying they were inferior to others and myself. I would probably need loyalty from these people in the future, so would continue to be courteous, and respectful.

"Do you want me to go in with you, or wait here?" Speedy asked as he pulled up to the building.

"Go in with me if you will. I could use some moral support," I replied.

The secretary escorted the both of us into a small room with several comfortable chairs around a table. Another person brought in a pitcher of liquid and cups. The council members filed in behind her and joined us at the table.

"Welcome King. How can we help you? What information do you seek?" one of them asked.

"I came to tell you of what my helpers advised should be done, and I'm inclined to agree. We would like to tour the country stopping long enough in each city to give a speech as a way of giving the people a chance to see who I am, and giving me a chance to become familiar with your culture. I would like to tell them the pride I have of being their king, and the support given to me by them, and their local council members.

I'm sure the discrimination we were subjected to on the other continent will not be as severe here. By giving them a chance to see and hear me, their reticence at accepting me would be reduced, or maybe even completely eliminated.

I would like your approval and blessing to do this. As you can tell now, that support and progress for this continent is foremost in my heart and efforts. If you would have any recommendations on how to go about it, or on what to say, that would be included as part of the planned agenda."

Then leaning back in the chair it seemed like an eternity for one of them to speak up about it. The chairman looked around the table at the others and made me wonder how the heck they were communicating or were they sitting there waiting for me to say something else. I was about to add to my comments when the chairman stood up and spoke.

"We like you wanting to become a kind and a giving King, and to become a valued member of this society. We will support you on this venture. Your expenses will be covered by the Council. The local authority will send us your bill, so you need not keep an account of it. If there are any concerns with what you will be saying or doing, then the local authorities will initiate an investigation. We will want a copy of the speech you are giving," he said.

"Oh, but I don't plan to use a prepared speech. I've already told you what I planned to say. If someone asks me a question on something, I would answer that honestly. I would expect a question on my nose, and that will be answered truthfully, and with a bit of humor. You know, something like, 'I have only one and you have two, so you see I got short-changed at the start'. Laughter always makes people relax and enjoy things more." was what I thought was a good reply. Then added. "The map has been studied. My driver knows the best way into and out of a city. I will get a competent helper to go with me to help on pronunciation, make sure I understand what is asked of me, and to say what I might have answered correctly. I would also need to explain that my vocal ability cannot make some of the sounds used in your speech. Part of my speech would also include a little demonstration on this problem of speech if that seems necessary. Just as I explained to you earlier that there are some sounds that my vocal cords will not make, they would have it explained to them the same way. Could make them enjoy seeing me in a more relaxed way, and realizing that being different can be fun rather than threatening.

"We wish you a very successful tour of our continent. We will monitor how well you are received in these cities, and if you need help will send somebody to assist you," the chairman said as they left the meeting.

Speedy and I got up to leave, and go back to our residence. As we drove along the way, the discussion with the Council members seemed to me it went very well. I even got to mention some talk of discrimination. Just didn't tell them how it would be presented. Speedy didn't catch it either, so there would be no explaining to him this day either.

When we got home, a phone call was made to John, the chairman of the local organization back home, to see what progress might have been done in this short time, and to see if they would know of someone to come work with me.

"Have you found anybody that would come stay here at the King house, and help set up my speech schedule?" I asked. I could hear John laughing as soon as he knew it was me.

"Yes, I sure have. I think you will be pleased to know that Cindy has agreed to come help. She has no work at this time, so this will fit in with her schedule very nicely. When would you like her to come?"

'As soon as it is convenient. I need to get started on this right away. Tell her not to worry about getting paid. It will all be taken care of. Thanks, John. I owe you one. The Council met with me this morning, and has approved the idea. They even thought it was a good idea." I said, and signed off.

The very next day Jane opened the door to let in two females carrying some kind of equipment. It was explained that this was language translator equipment. This would make it possible for me to speak in English, and the machine would translate to their language. Likewise, they could speak to me in their language, and the machine would translate it into English. But to do that the process would have to be developed further.

The equipment was set up, and ready to receive information. I had to say the words in English exactly as written. Then one of the females would repeat it after me, and again in the local language. By the time they were finished with it, the day was about gone. They expressed their appreciation of being so patient, packed up their equipment, and left.

The very next day the same two returned. The translator was ready. It was like a large medallion to hang around the neck. A small receiver about the size of an ear plug went into one ear. The plug was a wireless receiver matched to the same frequency as the translator hanging around my neck. It was an amazing machine. I can't say it any better than to call it wonderful. Such a blessing not to struggle with understanding what was being said. No more forcing my throat and vocal cords to try and make the right sounds of their language. It was impossible for me to make some of the sounds. I guess it would be similar to understanding the language of the Navajo code talkers used in World War Two. Then add the range of frequency to that and you can see how hard it would be.

The two ladies no sooner got out the door than Jane called me to take a message on the talkie. (I'm tired of this talkie stuff. From now on it's a phone.)

Cindy was waiting to talk with me and apparently had some questions.

"Greetings Daniel. When will you need me to be there?"

"It's not an emergency, but I would like you to come as soon as possible. We will have a lot of planning and work to do to get ready for the speaking tour," I said.

"It will be no problem for me to be there tomorrow. Do you expect me to stay overnight?" she asked.

"Yes, I will expect you to live here until the tour is completed if you can be away for a while," I replied.

"That doesn't seem quite proper. Are there assurances for my safety on this tour, and other places?"

"This is a very safe place. There are four full-time people to keep an eye on everything. One is a cook, one the housekeeper, a groundskeeper, and a driver. So no matter where we are, someone will always be there with us.

"Oh, I see." was her response.

I tried not to laugh too loud. "There are six bedrooms in this place, and every door has a lock. Every room has its own bathroom. The workers use several of the rooms of course, but I'm sure they will have one prepared very nicely for you. There is a very romantic garden setting though, and her laughter followed that comment. You can always pack up and leave too if that becomes a concern," I added, and quietly continued with my laughter just loud enough so she could hear.

"You aren't so darn hot that you couldn't be resisted by the most desperate of females Daniel, so quit your snickering. I wasn't one bit concerned about you anyway. There were situations there long before you came along, so I just wanted to make sure. Okay, I will be there late tomorrow," Cindy said, and now it was her turn to laugh just loud enough for me to hear.

I have to tell you that I was elated to hear Cindy's voice, and know she would be coming tomorrow. It has been very lonely not having friends my own age that can speak my language. I really liked Cindy as seemed to be a genuine friend. Now with this translator thing it will seem like we are all speaking the same.

The technology here is advanced much further than Earth. Electronics seems to be far ahead of Earth's electronic science. The big problem is the lack of minerals. Mining is very important, as there is not enough of most mineral ores that they can go through it like a little waste is no problem. So conservation is very necessary. I have no idea why there is such a shortage of iron ore. Copper and aluminum are in short supply as well as many other things I apparently know nothing about at this time.

With the development of matter-antimatter reaction producing power, then big power is no problem, but using this reaction is difficult to control in small doses, so solar power is the choice for most daily uses. When I learn more of the technology on all this I will try to give more thorough explanations later because at this point how it can be controlled stymies my brain. This material to produce such power cannot be found on this planet so they have to travel long distances to another planet to mine it. I guess that is what they were doing when I got stuck on their interplanetary ship.

Cindy got here. I was careful to give a polite reserved greeting. She was introduced to the staff and their duties were explained. I deliberately hadn't told her about my translator. She kept looking at me when I spoke. The speaker sounded a lot like my real voice speaking in her language, but it didn't take long before she knew. These translators have been used for quite a while to help people that had accidents to the throat area or those born with defects in the vocal cords.

"I'm so glad you got one of those. Now it won't be so hard to understand or figure out what you are trying to say. Your group at home told me they had a lot of trouble understanding what you were trying to say. You should go to their next meeting. Your talking in our language would give them much pleasure," she said.

The housekeeper Jane, gave Cindy a tour of the house and explained how it worked, and some of the schedule to keep things on time and running smoothly.

Jane helped carry Cindy's belongings into her room where the two spent the next couple of hours putting things away and talking. Sounded like they were getting along nicely. That ended when Cookie announced that dinner was ready. We all sat down at a table just like a family. Food was passed around the table for each to help themselves. I left the translator turned on which made the discussion much easier to understand and take part in the talking. To answer all the questions, it was necessary to give a full accounting of the trip to the other continent.

"Tell us about the match," Cookie said. "You have been back for several days now, and we haven't heard you say anything about it."

"The match next year will be in our country, so you can all watch how it goes yourself," I told them.

"No! No! You don't get off that easy. We would like to hear at least a little bit about it," Cindy spoke up.

"Okay if you really want to know. When we got to the fight area there was a big crowd waiting to see it. They all were cheering for their guy. No one to cheer for me. The King was introduced first. He got a big ovation. Then I was introduced to very polite applause. The fight didn't last long. The King badge was hung around my neck, and it was over," was the abbreviated version.

As long as we were on the subject that gave me the opportunity to see what their reaction would be to the discrimination shown there. The story may have been exaggerated a little, but I wanted to see what the reaction would be to the jeers, insults, and the discrimination shown to me. These four servants would need to be on my side if conflict came up. I was sure that Cindy would support me.

We talked a little more on the trip as Cookie got up to clear the table. Jane also got up to assist Cookie. Then Speedy and Blackie adjourned to go do whatever it was they do this time of day. With the clean-up job finished everybody went to their own room to call it a night.

With breakfast finished the next morning, I laid out my plans for Cindy to consider, and what she would be expected to do.

"When Speedy and I figure out which city to start with, and then what route to take to the next city your task is to contact some person in charge there to arrange a time and place for my speech. I'm sure they will know who to contact for publicity. Maybe they would have some good ideas on it and could arrange to have the city mayor or somebody important to welcome us and make an introduction to the crowd before I speak. Contacts could probably be done on the phone. Just be flexible in the scheduling in case the local person has some good suggestions on what they could do to make our visit more interesting and be seen as a favorable activity to the public.

"I would like the local dignitary to give a few comments on their new King. I will write what I want them to say. Then introduce you. From my experience on the other continent, the way my looks were met with disapproval some explanation of my one air intake instead of two should help ameliorate their reaction, and how they feel about my looks. Maybe a light-hearted comment would help. Something such as 'He may not be the prettiest king, but he is our king and we are keeping him. If you come up with something on your own that is okay with me.

I realize that is a lot to ask of you. Just keep in mind that this tour won't get done in just a few days. It may take several phone calls and even a trip to some cities to make arrangements. The four people from home should come to the first speech to see how it goes. Then our crew should get together after that first speech to analyze how it was accepted, and what modifications should be made in how it is presented."

"Do you expect me to stay with you that whole time? Am I expected to live here with you during that time? What can I expect in the way of compensation for all this work and time? "Cindy asked.

"Well, I don't know just yet. Guess I'll have to have another meeting with the Council. Of course they said at the last meeting to just go ahead with it, and turn the bills in to them to be taken care of. I would want you to stay with me though the whole tour. Tell you what. I will make the phone call to the Council to find out for sure on getting you paid," I replied.

"Okay, enough of this. I haven't seen much of this place. I'd like to see what the rest of the grounds surrounding the house looks like. I heard Speedy ask Blackie about something about a pond. Let us go look for the pond," Cindy suggested.

"You know, I haven't seen it either; been too darn busy to go look for it. Speedy must know where it is. I'll ask him," and started for the door.

"No, no. Where is your spirit of adventure? Let us just go look for it ourselves.
This place can't be so big we couldn't find it ourselves," Cindy grinned, and followed me out the door.

We spent a couple of lackadaisical hours walking out back of the house looking around at the plants and things growing there when we came across a narrow path. So we followed that for a way. Sure enough, it came to the edge of a pond. In size it looked like about one half of an Earth acre. We spotted a nice high-backed bench close to the edge of the pond, and headed for it to sit for a rest. Walking around to the front we were surprised to see Blackie sitting there half asleep. Gave him quite a start at the surprise of seeing us standing in front of him.

"I have to go now," Blackie said. "The car needs some attention. Just taking a break for a little while," and he got up to go. We had to laugh at that when he got out of earshot. We both knew he was taking a break alright, but it was obvious it was being enjoyed. We both sat down on the bench and leaned back to enjoy the nice gentle breeze blowing across the pond.

Cindy leaned over to give my hand a tender squeeze.

"I am so sorry for you Daniel. To be here on another planet where you have nothing. You have none of your old friends. You have no one with which to share your life. You have no means of earning a living. There is one thing you must do. Make the most of being the King. Endear yourself to my people, and in time you will be accepted as one of us. While you are the King you will be looked up to. You can go anywhere on this planet, and respect will be shown as a King. This will only be temporary, and once you are no longer the King then it is back to being just a foreigner in this land. So make the most of it. Cultivate your friendships now, and you will be able to make your way through life without undue hardship."

"Thank you for those encouraging words Cindy. Makes me feel worthwhile already. You are right. I will need to work on my people skills. I am a survivor, so being somebody that lives off of hand-outs is not in my future. Tell me one more thing.
Would it be possibly for one of the females on this planet to overlook my obvious difference, and live with me as a partner?" I asked.

"Not if they were wanting to have any offspring. Your genetics are different so it would not be possible. Your only hope would be to find an "inferior" for a partner. Even then that would take some getting used to your one nose right in the middle of everything. But if one were to find that you were kind and a gentle caring person, and they were having no luck in finding a partner then it could happen that a female could take an interest in you."

"Okay, I get the picture. About as much of a chance as a snowball in hell," was my smug reply.

"What is this snowball and hell? I don't know hell. Did you know this snowball person?" Cindy asked.

I had to break out laughing at that. Knew I shouldn't have used a term she wouldn't know, but sometimes I just have to express things like on Earth.

"Snow is the word we used on earth when precipitation in the air turns into flaky crystals when it's cold," I explained.

"Okay, then I know snowball. We have snow at both ends of our planet. Not much, but enough sometimes to make it hard to keep things from freezing. So now what is this hell," she asked.

"That is an imaginary place where bad people go to be punished. It is supposed to be very hot with fire.

"Then why do they go? Is it so hot they burn themselves?" she persisted.

"It is so hot they would burn up. But to be properly punished the fire would burn them for a long long time. But this in only an imaginary place that a man in a robe uses to scare people into believing a certain way."

I begged off trying to explain the concept of "hell" by saying we could talk about it at a later time.

With the discussion on the concept of heaven and hell over for now we went back to the house. That was one heck of a concept to be explained to somebody that had no previous exposure to religious ideas. I could understand why it would make no sense to Cindy or anybody else on this planet. Now that I was forced to think about it in a rational way, it was beginning to seem like it was the result of addle headed thinking right from the beginning.

If something that happened couldn't be explained as to why it happened, then it was blamed on forces that were made up by authorities that didn't know, but was expected to have the answer for it. How superstition and fallacy could eventually evolve into the various religions of today is amazing. And every one of them think that their religion is the only true religion. Then to want to kill for such a stupid belief is even more amazing. How stupid can Earth people be? Gullible, superstitious, easily lead astray, unreasonable followers, often cruel, nationalistic and territorial. To think I too was one of those people is now beginning to be inconceivable that I fell for the same fallacies. I was becoming confused if all the religious stuff was true, or stories from a shadowy past that been repeated and remodeled thru-out time that it was impossible to know for sure. Maybe this place is getting me confused on my beliefs.

Cindy got to work on the logistics of the first speech., and I got to work on the speech. The day was set, and the venue selected. The city leader picked to do the introductions was similar to the job of an Earth city mayor. As soon as the day was set, I called the local group to let them know, and ask them to be there for that first speech. Then Fred and Wilma were called. I wanted them to sit on the stage with the other important people. I hoped this would make them feel their effort in supporting me through the early learning stage was worthwhile.

The mayor did a nice introduction for Cindy. Then she did a good job of introducing me, and explaining my looks. When I walked out on the stage there were a few gasps at seeing me, but mostly it was pretty quiet with a few scattered cheers. Seemed it was going to take a while to get used to me. In all the time I had been here, there had not been a lot of comments on my hair. It had been cropped very short as a member of the wrestling team in college, but had been growing ever since. Without any barbers on this continent to keep it trimmed, things were getting pretty shaggy on top. I had been referred to as "one nose" or whatever their word is for nose. Lord knows what I'll be called when this head of hair gets long. Combine that with facial hair growing out all over my face, and my looks has been getting a bit scary for these people. Now that I'm going to be doing these talks in front of groups then Cindy will have to explain my appearance before I appear.

The first speech as King.

"Thank you for coming to my first speech as your King. As you know it was just a short time ago that I defeated the long-term King of that other Continent. Let me tell you that it was a hard-fought bout, but I was determined to bring back the badge of the King for you. I am here to support you any way that I can. My door is always open to any of you to listen to any and all requests. I will help you with any request if it is in my power to do so. But remember, a request may have to be approved by the Council.

I realize that I look different than you, but since my arrival I have become one of you. I admire your progress in technology. I admire your progress in agriculture, except for those fungus. When one hit the front of our car it splattered all over. Almost scared me right out of the seat, and made a mess of me as well. They sure do taste good though. This bit of humor was met with approval so decided to use it any future talks.

The progress in job training and hiring procedures need some additional attention. The present jobs pay good wages, and the managers all have good intentions, but there are so few "inferiors" employed, that it just is not right. Too many "inferiors" do not have a job because of the discrimination shown in hiring. So too many are going hungry because of no wages, or low wages. I have also personally seen the same kind of discrimination against "inferiors "on the streets and other public locations. This is wrong, and we need to adjust our thinking in society to avoid this problem. It should not be happening in a society as advanced as this one, which I now have to consider as mine as well as yours.

Thanks to your mayor for his help in welcoming us to this city, and thank you people for coming to support your King. Together we will work to make this one of the best cities in the whole world."

I never was much good at public speaking but hoped it went okay. As we left the stage, Cindy and the friends will offer their opinions when we get back to the house. Fred and Wilma congratulated me on the speech before heading to their car for the trip home. For the first time I expressed my love for them, and told them thanks for all they have done for me.

Speedy waited for the friends to get to their car, and then follow us back to the house. I was anxious to hear what Cindy thought of the speech, but she said nothing about it the entire way back to the house. I really wanted to know what she thought, but didn't want to appear to be too anxious to hear in case she thought it was a bad speech, or maybe even a disaster, as their wasn't a big enthusiastic display from the audience.

When we all got into the house and took a seat around the table, Cookie had a nice hot drink ready for us, and a very delicious dessert.

"Well, what did you think of the speech," I inquired to no one in particular. "Give me some feedback and how you thought it went. Bart spoke up immediately.

"I thought it was received well enough, but what the heck was that bit about hungry "inferiors". We maybe don't make as much as some of the regulars, and granted we do have a lot more trouble getting a job, but most of us don't go hungry for lack of food."

"Ya, I was wondering about that as well. What was the intentions with that comment? It didn't seem to mean much to anybody in the audience that I could tell," Sam said.

"Okay, let me explain. If we go along pretending nothing is wrong, we would never make any progress on discrimination. So if I just mention it in each speech that will at least get some of them thinking about it. Then as time goes on and I make more speeches the comments become more forceful, and louder.

With you guys going to difference cities, and organizing different chapters of our group, we can start a national organization to work for equality of "regulars" and "Inferiors". I know this could work as it worked on my other planet for John L. Lewis and Jimmy Hoffa. Except there they talked of cold and starving children going to bed each night because their parents had no money to buy food. In time the public came together to support them.

There will be resistance at the beginning, and in some cases maybe even push-back on what we are doing, but in the end with patience, and continual work the public will come around, and then we will even start getting support from them," I explained. "Does that seem like too much work for anybody? If it is, then you will never get this country over discrimination. With my different-looking face, you know that nobody would hire me. So I get discriminated on all counts. Every time I go into public, discrimination is on every bodies face. That discrimination is only because I look different. For me, that can never be eliminated. But for you, when you go out in public, nobody that looks at you can not see any difference, so you see no discrimination," I finished.

Lost in thought on it, nobody said anything for a few moments, then started to look at each other to see who would speak first. I would have liked to build a fire under them but didn't think they should be brow-beaten to get them to act. Up until that point, Cindy had remained quiet, then stood up to get our attention.

"Why do we have to go through all this organizing, and speaking? If all that has to change is the label on our official information then why not just concentrate on getting that label changed? See if we can get the national personnel record to just eliminate that category. Then make it the responsibility of the two people that are planning to be partners, to ask the other if they have clearance to reproduce. This information would be in the record, so certification could be obtained for each before the partnership was granted, and that would take care of that kind of discrimination". With that, Cindy sat down.

"But what about employment? Would the managers still have access to the official records on everybody?" Sam asked?

"That part of the record would have to be blocked, and only accessible to those wanting to establish a partnership," Cindy added.

This idea ruined my plans for fighting back on discrimination because of physical appearance, but everybody sitting at the table agreed with Cindy but me. Rather than have controversy this early in the relationship, I kept my mouth shut on it. This would be a good start anyway to changing things in people's attitude.

"Okay, this would be a good starting point. I shall continue the tour and the speeches. If you guys would continue with your visits to talk with managers and point out that "inferiors" can do the job just as well or even better than some regulars then we may gain enough support to convince the Council to modify the records, and who has access to them. So some of the things I will try to say and do will be things such as my face may look different from theirs, but it didn't prevent me from becoming the King. A proud and intelligent people would not want the country showing hostility to their King's appearance. After all, I fought for you to bring the King badge back home, and similar comments to instill some patriotism and pride. That line can be modified as we go from place to place.

Any other concerns we need to address today? If not then I will be in touch with you from time to time. It would be nice if you could attend all of the speeches to observe and let me know what modifications you think we should make in the content. Thanks for coming. Have a safe trip home."

The meeting broke up with a couple of the guys grabbing one last piece of Cookie's dessert. Only Cindy stayed behind to discuss the next city to visit.

"What do you think, Cindy? Am I so unpleasant to look at that the only thing that makes me acceptable is being the King? Sure, I am different, but not repulsive. Do you find it hard to look at me?" I asked.

"Well, not repulsive, but if you could get rid of that stuff on top of your head, and on your chin, it would be a lot easier for people to see you are not that different."

"Hair! Cindy's hair! That stuff on my head is called hair. Everybody, even the females have hair on their head on Earth. A nice head of hair is considered attractive to Earth people," I protested. "I know people are reluctant to touch me for fear of coming in contact with hair, but how about you? Are you afraid to touch me?"

"No of course not. I have touched you. If you would have noticed, it didn't bother me at all," she insisted.

"Then show me. Show me it doesn't bother you. Come hold my hand and walk with me. Let's go down to the pond and sit there to watch the sun go down, I insisted rising from the table and holding out my hand. "

"Sure! No problem for me to hold your hand," Cindy replied and rose to follow me outside.

She took my hand in hers. With those two thumbs, it always amazed me how strong of a grip she had. But there was no hard squeezing, just a nice tender pressure.

"So how does that feel to know you are holding hands with an alien, and not with one of your own handsome kind?" I chided.

"Don't be such a sarcastic jerk. Actually it's pretty nice. Your temperature is a little cooler than ours, so the coolness of your hand feels pretty good. Now shut up for a little while, and enjoy the nice early end of the day. I like to watch the sun disappear below the horizon, so come sit with me down at the pond and watch the day end," she smiled at me.

As soon as we sat down on the bench she turned loose of my hand, and leaned against the backrest. It dawned on me how much I had enjoyed sitting with a girl watching the moon rise back on Earth. I had to admit to myself that this was really nice. So holding her hand, and walking side by side made the world seem more secure than anything else I had felt or done since getting here. I shifted my weight as though to adjust myself, and leaned back again so that our shoulders just touched a bit. She did not recoil from that or even slightly move away so assumed that Cindy was accepting me more as a normal person.

This was very pleasant, but the sun was about to slip below the horizon so it was time to go. Then the feeling that the darkness would end the closeness I had been feeling, my thoughts turned to how it used to be. The moon coming up to cast shadows all around. The orb of light in the distant sky, and the warmth of a girl in my arms as we watched the sun and moon trade images of the evening shadows. But on this planet, there was no moon. Now on this planet when the sun was gone there was nothing of the light to make shadows. In the early darkness, the feelings of a lonely, unloved being was taking over. There was nothing here of romance in this land to make life worth living. How do you share your dreams when there is nobody with which to share a dream?

For a moment a thought maybe there could be something with Cindy, but it was nothing but the sarcasm in practically daring Cindy to hold my hand. It was nice, but I suspect she just wanted to make me feel better about my different looks. Thinking about that as we were walking back to the house, and not holding hands. I wondered if she would reach out to take my hand. But she didn't. Thinking about the problem my hair seemed to be for everybody, I posed the question.

"How would it look to everybody if I cut off all the hair on my head and face? Would I look more like the guys?" I asked. "I could do that you know."

"How would you do that? We have shears and scissors of course, but you have a lot on your face. That could be difficult to get to." she said.

"Yes, it would, but I could do it. Cookie has some pretty sharp knives in the kitchen. I would get it all lathered up with something, and then scrape it off with one of the knives," I explained. "But what do you think? Would it help to look more like one of the native people?"

"Well it probably would, but is that something you would want to do? The people know that you have hair, so why not save the trouble of trying to look like them? It wouldn't make any difference to me. I've gotten used to how you look. If you scrape all that hair off I would have to get used to what you look like all over again," Cindy laughed. "I would like to see how you scrape all that off with one of Cookie's knives, but I wouldn't want to hear you yelling how much it hurts when the knife cuts it off. You would really have to be careful doing it under your nose, or you could reduce your nose even further," she said trying to make light of it.

"Hey! Don't laugh about it. On Earth, we do that every day. But we have what are called razors. Some run on electricity, and some are battery-powered. We just turn it on, run it over our face, and the job is done. *It cuts the hair right off.* In earlier times before we had electric power, the whiskers, that's what face hair is called, was shaved off with a sharp blade. And yes it is true that occasionally you could get a cut on your face."

"I may give it a try anyway. If I cut off more nose, then the face will look more like yours, so what have I got to lose? You know, Fred and Wilma never once commented on my hair. I wonder why?"

"If I know anything about our government, they probably had instructions not to ask, or comment on your appearance. Or at least not until you became adjusted to our culture. If you couldn't fit in, then with your elimination it would have been pointless anyway to comment on it." Cindy suggested. "Your hair is such a mess the way it is now. Can't we just cut some of it off? At least enough so it doesn't look so shaggy.

"I was wondering about doing that myself, but I couldn't find any scissors, and I didn't want anybody watching, so just let it go until Wilma could help."

"I have some scissors. We may wear some form-fitting clothing with no frills, but they still need to be made and adjusted to fit the individual. That takes some equipment to do that you know. So you can use my scissors. Or better yet, I will help you so you don't cut off an ear or something," she chuckled. "First thing tomorrow morning after breakfast."

The next morning after a nice breakfast that Cookie provided, I was given instructions to go outside, take one of the chairs that lined the entryway, sit in it, and don't be telling her how to do the job.

Cindy came right behind me with some scissors, and a brush that looked like the one Cookie used in the kitchen to clean pans and stuff. I was about to say something, but thought better of it as I saw the expression on

Cindy's face. She could convey a meaning that meant "don't push it" quicker than anybody I ever knew.

She went to work on that mop of hair so fast I was in fear of what it would look like when she was finished, but again I kept my mouth shut. Apparently finished with my head, she moved to my front; pulled the sides of my beard in together, and held it in a fist under my chin. Shaking her head in disapproval, the scissors attacked my beard. I couldn't help noticing the way those two thumbs worked in unison. When she was satisfied, a step back to see how it looked, and then a couple of little snips here and there. It gave me quite a scare when she came at me again brandishing those scissors. One hand clamped down on the top of my head.

"Hold still," she demanded. "So you want to lose the rest of your nose? That hair sticking out there needs to be removed. It looks terrible. I'm just going to clip off the ends that stick out."

I froze in place. That was kind of a sensitive area anyway after the operation, and sure didn't want any more injury there, so admonished her to be careful. There was no recognition she even heard that, but the job was completed without any collateral damage.

Going back in the house to immediately find a mirror, it was a bit of a surprise to find she really did a nice job. It actually looked like a professional barber had done the work. It made me very proud of what she could accomplish and feeling a little sorry, I had tried to tell her how to do it.

I was so happy that the mop of hair now looked well-groomed, and even attractive by Earth standards. I was so pleased with it that my thanks didn't seem quite enough so reaching out I forgot about the standard way to do it, and instead of the hands on the shoulders, put both arms around her with a big hug of gratitude. She immediately tried to take a step back, but I hung on tight, pulled her closer, and finished with a big squeeze.

"What are you doing," she exclaimed in surprise.

"This is the way we do it on Earth, instead of the hands on the shoulders squeeze," I told her. "Sometimes to express thanks, and show appreciation we give somebody a hug. A hug is when you put your arms around them and embrace them with your arms," I explained.

"I thought that was a hug. Then you told me earlier that's an embrace. Which is it?"

"Actually it's about the same thing. You take them in your arms, pull them close to you and gently let them know you like what they have said or done," I hoped that would take care of the explanation.

"I didn't know that," she replied. "It was a sudden surprise move that I didn't know about. Seems kind of silly, but I like it. Do that again without the surprise, and show me how it is done."

I hadn't planned on giving her such a big hug, but the haircut made me very pleased, and the hug was spontaneous. Her request to be shown how it is done on Earth was also welcome, so the hug was repeated, but slower and more gentle than the first time.

"How is this action used on your planet? What is it conveyed to show, or do," she asked. "When is it used?"

"We use it when meeting someone we like, or even when they leave. We use it to greet relatives. It is often used to show appreciation like I just did. There are many reasons to use it. It is especially used with a person of the opposite sex that you want to establish a closer relationship with." I told her.

"I will have to think about this hug action," she said, and that ended it.

It may have ended it for her, but that last hug with that warm body pressed against mine brought back memories of the girl left behind. I tried not to think of those days at the university when we got back from a movie, or a party, and could sit in the car for hours with are arms around each other, and wish the night would never end. Deep depression would slowly override the thoughts of those wondrous times. How nice it might be to just end it all on this planet. Have done with this lonely life; have done with this King crap; have done with no girlfriend; have done with the speeches; have done with trying to get this planet organized: just have done with everything. Suicide is easy. Why not just have all this over and done with?

But when I think about how to do it, I'm too weak to carry it out. I may have won the fight, and I may be the King, but this King is too weak to win this fight with himself.

The next day brought a better outlook on life. Cookie was a cheerful greeter that morning. Cindy came in for breakfast, gave me a hug, and smiled.

"I think I did that hug right didn't I," she said.

"That was fine. Just the way it should be. Thank you for that," was all I could say, but it improved my outlook tremendously.

"I have the next speech date approved, and everything ready to go. It will be in two days. Have you got your talk ready, or will you use the same one as the last time?" Cindy asked.

"I'll use the last one, and add things that seem appropriate as we go along. The guys also need to be notified in case they want to be there. I will take care of that myself. Need to hear what they have been doing to promote the cause."

There was some surprise and gawking at my appearance as expected, but the speech went pretty good. My picture had been seen by most of the population by now, so the large turn-out was not only to hear the King's speech but to see what he looked like. The cheers were loud and enthusiastic when the speech was about their accomplishments, but much more subdued when it was about discrimination of "inferiors, or appearance" When the speech was over, many of them crowded closer to shake my hand as most had learned by now that was the way it was done where I came from. Their eyes though were focused on my head. One little kid wanted to touch my beard. I was glad Cindy had cut it into a nice rounded shape so bent down for him to feel it. Others got closer but held back just enough so they wouldn't have to touch me or shake my hand. The revulsion on their face at their difference from my face was obvious. My thoughts on this was to mentally just say to myself (to hell with it, I am the King. I could probably have you killed if I wanted.)

The translator was working nicely now as it was programmed to be smart enough to figure out with some usage what their language would be in

English. I wondered how many different languages there would be on both continents. Something like this translator would be extremely useful on Earth.

After the crowd cleared out, the guys from our group joined us in a meeting room. Cindy, efficient Cindy, had specified that a treat and drink be ready for us after the speech in the meeting room.

"How are things going with managers? Any progress on hiring practices?" I asked."

"Not much," Sam spoke up. "They listen to us, but there is a lot of resistance on making any changes on hiring, or pay. We inferiors always get less pay for doing the same job as anybody else."

"I think it's time for me to make a public statement. No, make that a King's edict. The disseminators will be notified tomorrow. It will be illegal for any manager to discriminate for any reason except for a person that cannot physically or mentally do the job. We will determine the punishment at a later date. So you guys get ready to put some pressure on those managers," I demanded. "Wait for the news to come out so the managers will know that it is now the law, and they will have to obey it, or be punished."

As we were getting ready to leave, the local authority introduced himself and asked if I would like a tour of the machine factory. Since I had never seen one of their operations it was readily agreed to. Cindy asked to be excused to take care of some other business, so I was the only one that went with him.

It was interesting watching the machines at work. There were just a few attendants keeping an eye on what the machines were doing.

"This is some of our latest technology. There is only one other factory with this advanced technology," he said. We watched for a while, then he led me to another room off to the side.

"This is something you have to see to believe. Several years ago when we were at war with the other continent there was a need for a special weapon. This was the result of our research. It is powered by a special crystal that concentrates the laser energy, then lets it go in a blast that can devastate whatever it hits. It hasn't been used for several years now, so it is no longer manufactured. Someday maybe we can take it out where I can show you how powerful it is,"

He handed me the gun to hold, and look it over. I would have liked to know how it all worked, but knew its technology was beyond my comprehension of laser physics. I thanked him for the tour, as he led me back to the conference room.

On the way back to the house, Cindy and I had a spirited discussion over what might happen. She agreed to notify the disseminators and set a time for the announcement. The day of the meeting and all the people that were responsible for getting the news to the public were assembled at the entryway where I would come out and give the edict.

"Henceforth there will be no discrimination in the hiring of "inferiors", no discrimination for appearance, or the sex of the applicant. Pay will be the same for all who do the same work. The only exception would be for a person unable physically or mentally to do the job for which he or she applied. This by order of the King."

The information began flowing out to the public immediately. As I had expected, but also dreaded, the Council called me in for a meeting the very next morning. I knew this was coming, so was prepared for what might happen. I asked Blackie to not wear his uniform, and that he and Blackie dress as closely alike as they could. When both were ready we headed for the Council office.

The head guy was ready. With his notes in front of him, he set forth what my duties were supposed to be. I listened but said nothing. He made one final comment.

"You cannot make new laws. Laws are made by us," he finished.

"I am the King. A king sets rules and regulations," I replied. What is it you object to on what is said? Do you agree that everybody should be treated fairly?"

"Of course," was his answer.

"Do you agree that people should not be judged by the looks they were born with?" I asked.

"Of course," he agreed.

"If two people apply for the same job would you agree it should go to the one that is best qualified, and capable of doing it?" I questioned again.

"Of course, but what are you getting at? End this silly questioning, and say what you want." He obviously wanted this to end, but I would not capitulate until my part of this was done.

"Would you agree that it shouldn't make any different what sex the applicant was if either one was qualified to do that job,? I continued.

"Yes I agree, I agree," and it was obvious he was getting a little agitated with me.

It looked like it was time to play the trump card, so Blackie and Speedie were asked to step forward.

"With these two gentlemen standing before you, can you tell me which one should be given the job if they are equally qualified, and they both have demonstrated that they can do the job with equal efficiency?" I asked.

"No, I can't tell who should get it. What is the purpose of this nonsense?" he snapped.

"Would you look at the official list of names in the national register to see if there is anything there that would be a disqualification?" I continued the questioning.

"Yes, I would. Of course, I would."

"What would disqualify either one of them?" I asked.

"A negative report on past jobs."

"What else?" I persisted. "Would it make any difference if he had a child at home?"

"No, it wouldn't make any difference if he had none or many. Okay, that is enough. Ether get to the point, or get out. He snarled.

"That is exactly the point. If your national register had them listed as inferior because of a minor genetic default, neither one of them would get the job. That notation is meaningless for job qualifications. It is one of the most insidious causes of discrimination I've seen. That notation must be removed from the register.

That is precisely why I gave the edict on it yesterday. I am the king, and that edict will stand whether you approve of it or not. One other item. I am the King, and if you talk to this King like that again it will be you that will be out," and turning on my heel, motioned that Speedie and Blackie follow. There was dead silence behind us as we left.

I couldn't back down now and would just have to wait to see what they would do about it. So with Cindy getting me the connections to the registered office, the call was made to them. The office manager was instructed to eliminate that category from the files. It had to be explained that the Council had been advised of this change by order of the King, and this change was to be made immediately. The Council would take a few days to debate the issue, but by the time they got around to making a decision, the change would have already been done.

To cause them some confusion, and to ensure they would have to debate the situation for some time before finding out it was too late. The group was called to come with as many supporters as they could get.

That next morning it took quite a bit of explaining what I wanted them to do. Using a page from the" occupy wall street" protest movement on Earth trying to make changes, my explanation of what they would do was soon met with enthusiasm. Nothing like this had ever happened that anybody could remember, so this new experience was a novel adventure for them.

As they headed to the Council office, Cindy called the media reporters to say there was some kind of protest activity taking place at the Council. I couldn't go along or be seen as supporting this protest because it might bring negative publicity on the King. It would have been interesting to see what the reaction would be by the Council members as it was happening.

The group dispersed after their demonstration was over, except for Sam who returned to the house with his report. There was no force used by either side; no problems getting into the chambers; no attempt to eject them; the Council members just sat there and listened to the demands about discrimination. And getting the register changed. The media people recorded all that happened, and everybody left.

I was pleased to hear there was no violence, but with not much of a reaction wondered it it had any effect. Sam convinced me that this was a normal reaction from a peaceable people. Here they present their case. It would be taken into account, and a decision rendered. Whatever the decision it was simply accepted, and obeyed. My final instructions for Sam was to keep an eye on job openings and encourage "inferiors" to apply. We would soon see if the register had eliminated that category of information. The council members would be in for a surprised to find out that the provision in the register had already been removed. I hoped there would be no push-back on the King once they found out who was responsible.

I had been busy as a King traveling from city to city giving speeches, and visiting factories, and places of interest. My knowledge of this planet, and the people on it from all the places I had been, to convince me that there was not much diversity. This may be a small planet, but on Earth, there was lots of variations in the population even in just one country, let alone the whole world. The translator had been invaluable. It has been with me wherever I go. Trying to understand and speak their language is impossible, for many of the sounds cannot be made by my vocal cords. Thankfully this thing seems to be self-adjusting as it "learns" how some of my speech affects meanings.

The year was almost over, and my reign as King would end if I did not fight in the next tournament. Notices had gone out. Times and place of locals matches were published. I did not have to participate in the tournament. As the King, the only match for me was to take on the final challenger of the tournament. I went back to the old routine of conditioning. Cookies fantastic meals made it hard to resist eating so much, and as a result, some weight had been added to my frame.

I AM THE ALIEN

Cindy understands what I'm doing, and why, but refuses to take part in it. I can't even entice her to accompany me on a nice easy run. Lifting weights, and doing stretches seem to be a total waste of time to her. There are times though when I notice her looking at my body with what I think is admiration. Maybe I'm mistaking contempt for admiration. It's hard to tell what some of the facial expressions mean.

The day came when the tournament was complete. The winner came from my old section of the country where Fred and Wilma lived. Came time for the match with me, and the opponent decided he couldn't beat me, and forfeited. A lot of people were coming to see the match and would be very disappointed if none took place.

I ask to meet the competition to thank him for withdrawing, but I had other plans. I wanted a sparring partner and he was probably the best that could be had. Explaining to him that we would go through with the fight. He would do his best, but I would pull my punches, and not hurt him. We would at least give them a bit of a show. When it was enough we would get in a clinch, and I would say stop. Then I would pretend to hit him with a good one. He would drop to the ground, and that would end it. This seemed like a good idea to him as he would not appear to be afraid of a fight with me.

The match was arranged. Lots of people came and gathered around the square. The announcer started the ceremonies. For a few moments, we made the fight look pretty good. As per instructions he was trying to get in some hood hits. When I thought that was good enough of a show, I stepped in to get in a clinch and said stop. Stepping back, then making an exaggerated swing pretended to hit him with a powerful shot.

He went down but got right back up to continue the fight. I thought maybe he just temporarily forgot, so stepped in again to say stop. Then step back, make the exaggerated swing, and he was supposed to go down. Again he popped right back up. I didn't know what the heck he was doing, so decided to end it myself. This time I fainted with a left, then a short sharp right hook to the jaw. He went down like a sack of potatoes, and this time he did not get up. I hated to do it, but the plan was not being followed, and there was no clue as to what he was going to do.

The crowd went home satisfied they had seen the fight. I did not meet with the loser afterward either. By ignoring him that seemed like enough disrespect to show I was not happy with his performance.

Cindy had watched the fight but showed no reaction when it was over. Just said congratulations, and went on her way. Late that afternoon though she came to me and asked.

"Will you walk down to the pond with me? I would like to talk with you."

"Ya, sure. I ways enjoy sitting with you at the pond," I replied. So I got up, and joined her at the door to go.

Starting down the path she reached over, and took my hand. We walked together in silence for a ways. Ever since that first time when she was challenged to hold my hand, every time we went to the pond she reached out and took my hand. That was the only time she would hold my hand, but she always did during that walk. I have no idea why unless she thought it was her duty. Why wouldn't she do it at other times? Sitting at the bench I turned to face her and ask.

"Okay, what did you want to talk about?"

"Nothing really important. I've been extremely busy this past year getting all your activities scheduled, and the things done that were required. I'm getting a little homesick. I want to go home for a while to see my parents. You won't need me for a while anyway. You are too busy training for the King match coming up before long. You've been all over the continent giving speeches, so that can stop too. Cookie and Jane can take care of anything you might need."

"Oh! I was wondering why you seemed to be down in the dumps lately. Couldn't figure out why you seemed unhappy. Certainly, you can go home for a while. Forgive me for not thinking about what you wanted or needed to do. Of Course, you should go see your parents. Speedy will drive you there. When you are ready to come back just call, and he will come to get you."

"What is this dump thing you said I was in? The translator doesn't seem to know," she asked.

"I used a word there that is just an expression of how someone feels at the moment." It's not important, but I had to smile to myself at the double meaning some of the English words can have.

I suddenly felt a little guilty at not considering her feelings. I wondered if a real King ever thought about the people that served him. Then thoughts turned to the four servants. Cookie, Jane, Blackie, and Speedie had never asked for anything for themselves. Maybe I should ask them. But my thoughts turned to the perks of being King. That is their job. Take care of the King. Do what he wants. Be subservient in every way. That is their life, so no need to worry about them.

"You use so many words that the translator doesn't do very good at translation. So if I have to ask so often I am sorry for that inconvenience Just as soon as I call my mother to see when it is a good time to come I will let you know," she said.

Negotiations were underway with the other continent to set a date for the King Match. It is always held on the continent where the present King is located, so this time it will be on our continent. Preparations for the event were rather elaborate. Colorful decorations, and big placards were placed everywhere along the route to the capitol. I decided to make one request to my people to go out to the media. The jeers and discrimination I felt when the match was held on the other continent, made me feel very bad, and I wanted to see no repeat of that on my continent.

The delegation would be met with respect. A welcome committee would meet the boat with signs of welcome, then escort to lodgings that were obviously of first-class quality. The crowd at the match would be advised how to treat the competitor who was the defeated King of last year's bout that was here to reclaim the badge.

I thought it was my duty to personally welcome the defeated King, so made a visit to his lodgings. I made the mistake of telling him how much stronger, and faster I was. I didn't want to seriously hurt him in front of all these people, so he should quit early in the fight. This really backfired. He was insulted by this. He took that opportunity to call me ugly, a disgrace to all the people on my continent, and he would not only win but beat me into the ground; then step on my ugly face as he left the square. To prohibit further escalation of this attitude further, a hasty exit was made with obscene comments following me out of there.

The people of my continent poured into the arena by the thousands. I was proud of the reception they gave the past King. I was properly coached on the formalities that preceded the fight. When we faced each other ready to start the duel, the contempt on his face brought my blood to a boil. He would be shown who was the superior in this fight. Now I was prepared to hurt him with no mercy shown.

I don't remember much of the preliminaries now, but I remember every blow of the fight. He came at me with a snarl on his lips and started swinging. It was easy to avoid being hit with his slower movements than mine. I could have knocked him silly with one blow to the side of his jaw. But I chose his chest. A short sharp right into the left side sent him reeling. That really hurt. He came back at me with his arm lowered a bit to protect that side from another blow. So I feinted with my right to that side, and as he flinched to turn that side slightly away my left hand barreled into his other side.

Staggering to stay on his feet he wasn't in such a hurry to get at me this time. Pain showed on his face, but I didn't care, so now he would get it on both of his air holes. So a good shot on his left side of his face, and another on the right side put him down on one knee. I could see he was hurting as his face was contorted with pain as he forced himself to stand up. He knew he was beaten, but pride would not let him say "enough". As I approached him for one last blow the temptation was too much, so I got in one last comment. I don't know if he understood my attempt to say it, but I said, "Now who has the ugly face. Watch whose face gets stepped on as I walk out of here."

I doubt if he felt anything as my right fist crashed into the side of his face. His head snapped around from the blow. He was out stone cold in an instant before hitting the ground. He lay there not moving, and I almost carried through with stepping on his face. I went to the table that held the badge, picked it up, and marched out of the arena. This left the officials stunned, but I really didn't care. I knew he was dead. The urge to get away from there was the most important thing for me to do at the moment.

I knew he was dead and didn't really care. My experience with those inhabitants from the other continent had been so negative it affected how I viewed them. They weren't human, just an animal of some kind. The whole continent could be wiped out, and I wouldn't give a dam, I was supposed to be the King of this whole planet. The Council recommended a tour on their continent to show unity and goodwill should take place similar to the tour I was doing here, but now I didn't want to give them the satisfaction of hosting a tour for a King.

It was a comfort to get back to the house. It felt safe here with the servants and Cindy. A good workout did much to relieve the stress and my attitude. Sitting at the table later that day, munching on one of Cookie's treats, and a hot drink, I was feeling better when Cindy joined me. Cookie and Jane also joined us, so the four of us sat there feeding our faces, and sipping away at the drink.

"What law or rule would you like to see changed?" I asked of nobody in particular.

"You know, I have thought for some time now, that even with a very minor genetic defect, it would have been nice to have had a family of my own," Cindy said.

Then it came out. Both Cookie and Jane had been classified "inferior" too, and both resented it. When you have been sterilized as a child, it was impossible to get permission to have it reversed.

"You are the King. Can't you do something to get that requirement changed, or modified somehow so those of us with such a minor defect could still have a baby of our own? I would make a good mother. I know I would," Jane complained.

"Tell you what I'll do. Tomorrow if Speedy knows where it is then he can drive Cindy and me to the registrar's office to find out what things disqualify a person from having a baby," I said.

The next day, Speedie located the address of the registrar and took Cindy and myself to the office. The official there was very reluctant to even talk to us let alone disclose any information. As the King, I demanded to see what list of defects was considered unacceptable to be a father or mother.

The more I thought about it, the angrier I became that a government could have that much control over your life. It may be for the benefit of maintaining or even improving the general health of the population. But it seems to me that it narrows the focus of what the inhabitants develop into after several generations down the road. Any mutation that would lead to advancement would be dead on arrival since it could not be passed on if the law says any genetic variation is not allowed. I don't know why I should care anyway about these non-human people, but it just ticks me off on the excessive control over their lives.

I have nothing else to live for anyway, so may as well work on changing things for the benefit of those people labeled as "inferior". So come out with my second edict right there in the registrar's office.

"By rule of the King There is to be no mandatory genetic testing unless an individual asks for it. No person will be sterilized at any age unless asked for by that individual."

I expected a call from the Council again, and sure enough the next day the call came to appear before the Council immediately. Not wanting to get anybody else into trouble, it was only Speedy and myself that went to the meeting. Before the Council members had a chance to start questioning me on why the edict, and that I had no authority to do that, I repeated my edict, and this edict by the King would stand.

"No. That edict will not stand. We will be contacting the registrar's office today, and that law will remain in force. You are not to make any changes in a law on your own. You may be the King, but there are limitations on what you can and cannot do," they informed me.

"Don't cross me on this. The genetics law violates the rights of your people. They are free to choose for themselves whether they are capable of becoming parents. They are also smart enough to ask for a genetic test, and act according to the result. Minor problems are seldom a concern for a newborn. The people will not stand for this anymore," I insisted, and left the room.

This confrontation with the Council had me worried about overdoing it and getting them really upset. This King thing was a very comfy position to be in, and I sure didn't want to lose it. The food was good. The house was excellent, and the facilities were good, the servants were fantastic. But what the hell? If they kicked me out of the house and took away the title, what difference did it make? I had nothing to live for on this planet anyway. If they found me guilty of treason and sent me to be executed then I'd have it over and done with. That wouldn't be the first King that went to an executioner, at least on Earth. Life went on anyway, and the King went down in the history books. Maybe that should be my legacy too. Written up in the history books of this planet.

The situation was discussed with Cindy first. She thought it would probably be accepted by the general public, but not by the Council. Deciding to follow my own thought on it, the media people were called for an announcement where I repeated the edict on genetic testing and sterilization.

When this news release went out to the public there was a tremendous amount of conversation about it. Most of the population would support the idea. It seemed that it was mostly the authority figures that objected to it.

The next day the Council wanted to talk with me again. Speedy took me there of course. Again I was taken to task for an announcement not approved by the Council.

"We will remove you as the King if you go against this Council one more time. You have no authority here to make rules, or announce rules." they reminded me.

"I won the King badge on my own. There was not one of you that helped me do it. My term is up when I lose the King fight and not before," I countered. "Until then the rules will be set by me. Do I make myself clear? It will be the Council that will be replaced," I bluffed.

Not wanting to get into a further argument, Speedy and I left. On the way back home my mind was spinning over what would happen if they did try to remove me as King. What should my next move be if they counter-mended my orders to the registrar? If they did, then some kind of action would have to be taken, or I would lose support on getting any changes made. I would wait and see.

A few days later nothing had been changed, but that afternoon a group of four uniformed guys paid a little visit. There was no violence, but it was clear that there should be no interference with the Council's business, and that could be no change in the law without the approval of the Council. I listened politely but said nothing in the way of agreement. The message however was clear. Either butt out or risk action by the enforcement of whatever this group would do. Further study on the enforcement by the establishment would be my immediate concern. Have to check their power, where the power comes from, and who is the one giving orders.

I turned to Cindy and Speedy for information. Seems the president of the Council controls the police force, and orders go down the line from the president to district heads, then to the local level. This information told me that to control anything on this continent I would have to control the Council.

I had Speedy take me to the factory where the laser gun was developed. It was during the second speech where we went on a tour of the factory, and were shown the one gun that had been developed, and saved just in case it was needed some time in the future. I figured this was the future, so demanded that the owners and engineers make six of these laser guns. It took a demand by the authority of the King to convince them that these guns must be made. It was to be kept secret. Not even the Council members would be told of this project.

When the guns were ready, Speedy and I went to pick them up at the factory. Going to a room that was completely closed off from the main building we got a lesson on how to use them. These things would blast the target to smithereens and barely make a sound. The engineer was sworn to secrecy, and assured that the guns would be well controlled as long as they were in his laboratory setting. It took a bit of negotiating to let us take one to test what it could do and to experiment on how best to use it.

I had no idea of how it was possible to use matter and antimatter to produce such power. How would atoms of the matter be held in an area to prevent action, and likewise how could the antimatter be held in an area to prevent action there? Then bringing atoms of each together in the same chamber to get the explosion in a gun was something my limited Scientology deficient couldn't understand.

At this point, I had no intention of using them, but if threatened there were no inhibitions about using them to preserve and protect my own life. Fearing that there may be intentions of getting me out of the way, the group of friends were called on for a meeting. I would go there instead of having them come to my place as there was some concern over a listening device may be planted in the house. The day of the meeting it was Speedy of course that took us there. When all were there and seated around a table, and the small talk was out of the way it was time to explain why this meeting was needed.

"The Council is seriously thinking about trying to remove me as King. There has been a lot of effort on my part fighting to roll back some of the oppressive rules that have kept you down. The Council does not like what has been said, or what has been done to give you more freedom. You have been discriminated against in marriage, family, love, and a fair chance at employment. If this effort is to continue you may be called on to put your life on the line to defend your rights.

If the Council comes after me, I will ask that you be ready to take the place of those members. Start now to learn as much as you can about all the laws and rules that govern this country so when you are called to replace the Council members it will be with knowledgeable replacements," I said. "Are there any questions?"

"How will you be able to replace them?" Sam asked."

"Only two possibilities. Ask them to leave, or forcibly take them out. If it has to be by force, then I will personally do it," I answered.

I guess nobody wanted to speculate on how that would be done so we all sat in silent thought for a few moments, until I stood up, and announced it was time for me to go. Speedy had been listening to all that was said, but remained silent until we were in the car heading back.

"You could be in big trouble boss if it doesn't go right," Speedy said.

"Ya, I know, but somebody has to get their feet off the necks of the people," I responded. "But who else has the guts to do it? How do you get to be on the Council? What is the process?"

"There is no process. I don't know how the original members got started. The only thing I know is that the Council members pick their replacements when one of them steps down or dies. We have no say about who gets on," Speedy explained.

As I suspected, this government was a dictatorship. Maybe not by one, but by committee. The law was set with no hearings or participation with votes, or even with an expression of opinions.

"This means the only way to get rid of a Council member then is to kill him," I added to the conversation.

"Then the other members would have the killer arrested and he would be killed. If somebody gets too far out of line that is exactly what the Council does. They just have him killed. Have you noticed that we have no places of incarceration here? A violator of the law is simply eliminated rather than be restrained. Speedy explained. "Is that why you got the weapons? To protect yourself, or to kill the Council members?"

"Maybe both. I don't know yet. We will just have to see how it plays out. Hearing what you just said, I will have no reservations about either or both. In any event, we must be ready," I answered.

Speedy didn't seem to have any love for the Council. It seems he had been interested in not making waves of any kind as this was the best way to get through life staying out of trouble or conflict. Now it became clear why he never said anything about any aspect of life on this planet. So Speedy may even be supportive of what actions I take against the Council. Hearing how cheaply they valued life made me feel I was justified in my low opinion of these non-human beings. They make a great show of showing love of life, but in reality a death is not that big a deal.

We arrived back at the house. I thanked Speedy for the discussion and the valuable information on the Council. Tomorrow would be a visit with Blackie to see if he felt the same way Speedy did on the way things are done here. It would be good to have both on my side if things get rough for me.

The next day Blackie was called into the house for some talk. I complimented him on what a nice job he was doing with the grounds around the house. With one of Cookie's delicious desserts and a cup of hot berry juice in front of us, I gradually got into the problem of the Council. By the time our visit ended, Blackie had convinced me he would side with Speedy and my group on whatever decisions were made if it would be for the benefit of society.

Cookie had been listening from the next room, and wanted to be in on any action taken by my group. She and Jane had been keeping this place in good condition for a good many years for little credit and low pay. Cookie didn't think Jane would be too enthused but could be persuaded to join us if there was no harm to her.

The next step was for Speedy to drive us to the office that took in all the funds to run the country where I got a lesson on how the funds were dispersed. It was kept casual, no tough questions, a complimentary visit. I wanted it to come off as a visit by the King just wishing everybody a thank you for doing a nice job.

Nothing was done for quite a few days, No changes were pushed hard enough to cause any conflict, so things were going along pretty nicely until the parents of a young female was sent a notice to bring her in for genetic testing, and to be sterilized until she came of age.

The parents had been following the media on our edict of no more genetic testing and sterilization unless it was approved by them. So they decided not to do this with their progeny. After receiving several notices, the authorities came to their house. A date was set, and if they did not show up then the law would take care of it for the final result.

Sam notified me of the situation, so taking three of the laser guns that the engineer finally let us have, Speedy and I headed for the place where a confrontation would probably take place. We would try to avoid any violence, but just in case it was needed, the gun would definitely be used.

It was worse than I expected. The enforcers didn't waste any time. They showed up with six of them; parked the vehicle in the roadway; marched up to the house, and demanded that the parents come out, and bring the girl with them Instead of the parents going out, I had Harry go out to tell them nobody would be coming, so they could just go back.

The guy in front removed a gun from its holster and the laser put a nasty wound on Harry. That did it for me, so my own lasers were put to instant use. It was such a surprise that we even had the lasers that all six were blown away before they had a chance to react. The bodies were loaded back into their vehicles, driven back to their office, parked in front, left there, and we all went back.

I knew right then that would be my end if nothing was done, so on the way back to our house at the capital, Speedy and I discussed what to do. Speedy was sure tomorrow would be my last day alive, so somehow plans to get over to the other continent would be my best chance of escape. With some time to think about it that evening, the odds of even getting to a boat looked pretty slim, so I would have to do what I wanted to do for all of the past year. It was obvious these people could give a damn about a life, so why should I. They weren't human. Seemed more like an animal to me anyway.

I would just do an "Al Capone" on the damn Council. And I did. The next morning Speedy took us to the Council offices. They were already in the meeting room trying to decide what to do about it. Just as we were about to enter, their decision had been made. The King must be eliminated. Stepping through the door, I opened fire with my laser. All of them had no chance but to die where they sat. They thought they controlled everything, but now they controlled nothing. Hurrying back to the house, a quick call to Sam to bring the four most trusted friends, and come now. Don't waste any time. Just get here quickly. The revolution has begun.

Waiting for the group to get here, a speech had to go out to the media soon to calm the populace. They would need to be assured that nothing would happen to them. They were safe to go about their business as usual. This would be my speech to them.

"Citizens of our continent. There has been a change at the National Council. Nothing will change in your life. It is perfectly safe to go about your business as usual. The only change is in the make-up of the members of the Council. The king is in full control and wants to assure you that he only wants the best for you. He fought to bring the king badge to this continent for you, and he will fight to protect your rights."

This speech was delivered to all of the media sources. Since they knew nothing of what had happened just yet, nothing would be done to comment on what happened or why. In time it would become clear that a coup had taken place, and all that could be done was wait and see what came of it.

When the five guys got here I explained what had happened. They were to decide among themselves who would be best in which position to take the lead in a discussion of what should be changed, or if any changes were needed. They were to take up residence in the council building. One of the most important positions was the treasury. A salary would be determined for each based on the salary of the old members.

District authorities would be called in and informed about what happened, and if their loyalties were with the new council then continued employment would be considered with a raise in pay. If the loyalty was not with the new council then they were immediately terminated and disappeared from society. This was hard to do, but necessary for continuity. Saddam Hussein didn't hesitate to use this method of control for years as a very effective way of keeping detractors in line.

This all seemed like a strange twist of things. The outcasts or "inferiors" were now the ones in charge of everything. It was an easy task in having the new Council members pass a law making the King the supreme arbiter and maker of new laws either by decree or with the recommendation of the Council. This gave me power that could only be dreamed of. Now the world, or this continent anyway, was mine, and I was running it as I saw fit. Taking a life didn't seem to mean much in this society, so if anyone got in my way they would just be eliminated. My study of history on how it was done by most of the dictators though out the centuries on Earth proved to me that you could control any group of people with those methods. If Earth people find this account of my life here they will know what I'm saying. If non-Earth beings read it then it won't make any difference what I say here. If you understand the history of Joseph Stalin then you will understand the path I am taking to gain and keep control of this country.

It took some time for the population to understand what had really taken place on this continent. Little had changed in their lives, so for most, it didn't seem relevant at the time. Many were relieved that the required dreaded genetic testing and gene mapping had ended. Some voluntarily asked for and received the genetic evaluation. The registrar's office only gave out the information if it was asked for, and permitted by the ones involved.

Employment was now being done only by qualifications to do the job. Pay for work done was equal for all, and most of the populations seemed to be satisfied with their life. The attention of the genetic sciences was being focused more on improving plant species in both nutrition and production. I had taken a hard line with these people, but it seemed to be working very nicely.

My new powers as the King was beginning to improve my feeling of accomplishment, and give purpose to existence on this dinky little planet. But there was one thing I missed more than anything else, and that was to have a wife. My sex life had been non-existent ever since the capture on Earth and the looks of these people was revolting enough to prevent enough of an attraction to even think seriously about a partnership. I had teased Cindy in the earlier days about being reluctant to even touch me, but now the thought of more intimacy was whirling around in my mind almost every day now. These people all looked alike. Those genetic tests, and the elimination of those with what was considered inferior traits were not allowed to reproduce, had narrowed the gene pool so much that now you couldn't tell one from another.

I needed to get out of here, and somehow get back to Earth where a suitable lady would be waiting for me. Somewhere in my old city when I was enrolled at the university, I remember several that would make a great wife. At this stage of the game, most anything would be acceptable if it would get me back to Earth.

The decision was made to schedule a meeting with the space scientists that built the ship that captured me on Earth. If they could go past Earth once, then why not again. It was going well here, but I wanted to go home. That meeting was disappointing. They could go that far again, but the problem would be finding the right planet among all the others out there. It was left

I AM THE ALIEN

that they would continue going through the log of their travels until the right planet was found. If they stopped there to get a water supply, then it should show up some where in the log book. They were informed to keep searching for I would be back periodically to check on it.

The days were going by at a rapid rate now. Things seemed to be going pretty good throughout the continent. There were only sporadic objections to any new rules or regulations. Checking with the treasury office the funds were coming in at a steady rate. I hadn't checked with the new council for a while, so decided to pay them an unscheduled visit to say hello to old friends, and see how it was going with their duties.

Each one gave a brief report on what was being accomplished in their department, when John asked if they could have another raise in pay.

"What do you mean another raise in pay?" I asked. "Have you already given yourselves an increase?"

"Well, yes we have." Sam replied. "It is deserved. There was a lot of stuff that had to be worked on to put in order. All of us have put in extra time, and work."

"Okay, no problem. I just hoped you would let me know when you decided to modify anything. What about my funds? Do I get an increase in my income as well?" I asked.

"No, you don't need an increase. All you do is turn in your expenses, and it will be taken care of by the treasury. For your personal amount I thought you knew that you could just require the bookkeeper at the office to credit your account with what you wanted," Sam said. "By the way, it's about time for the next King competition you know. Do you want us to handle the arrangements?"

"I've been thinking about that. I believe we should not do it. We should declare that the fight is outdated, and has outlived its usefulness. I shall just remain as the King. If the other continent wants a king then let them get their own. I don't seem to have any control over their continent anyway, and they aren't going to tell me how we should run this continent," I said. "Okay, keep up the good work," and I left.

But I went directly to the treasury office to meet with the bookkeeper. I didn't want to meet with the head guy as he was one of us, and any shenanigans would have to go through him. So the instructions were made so that there was no mistaking what I wanted.

"From now on when there is a deduction made that is not an expense in running the operations of the continent, I want a copy sent to me. Nobody is to know of this except you and me. Am I clear on that?" I said in a serious tone.

"Yes sir. You have my assurance that it will be done exactly as you specify. Any changes you want done just let me know."

There was probably no need to be concerned about it, but better to keep up on everything that the Council is doing, and let them know they are being monitored.

The King fight popped into my mind on the way back to the house. That fight was a bunch of nonsense anyway and didn't need to be done again. I didn't want to do it, but I didn't want to lose the present position of King. It gave me something to do, and provided me with the easiest life here. House provided, travel provided, cooking and cleaning provided. What more could I want? If I had a woman in my life it would be perfect. But that is one thing I didn't have knawed at my mind every day. The workout getting in condition for the fight, I had to admit to myself, kept me from punching holes in the walls to relieve the tension sometimes. When sitting at the kitchen table, it didn't take long for Cookie to serve something she always had that was ready to eat. Cookie didn't say a lot, but always seemed to be knowledgeable about what was going on. So I proposed the idea to see what her reaction would be to calling off the fight.

"Cookie, I'm thinking of ending the competition to be the King. What if an announcement was made that I was going to be the permanent King of this continent, and the other continent could have their own King? There would be no need for the fight. What do you think of that idea?"

"Excuse me, but I think that would be a big mistake. First of all, that annual fight replaced the real fight we had years ago when a lot of people were killed, and buildings destroyed. It is the one big competition we now have. There is no other competition because it would give rise to increasing competition and nationalistic feelings until war would start again. When we regain the King lost in the fight it is like the country is reborn again. If there were no King fights then the countries would not have that feeling of a new start. That is what we were taught as kids, and I still believe it. No, I would not be in favor of such a change. Please do not do that to this country," Cookie pleaded.

"Okay, I just wanted to see what you thought about that change. I will ask Cindy and also Speedy what they think," I said in a gentler tone.

During the next few days, I ran the idea by both Speedy and Blackie. They both agreed with Cookie. A king should earn it, and not just be one because he says he is King. That was enough for me that I didn't need Cindy's opinion on it. So I would go through with one more fight, and then that was it. I would become a permanent King, or else just quit. If the scientists could find the route back to Earth, then I'm out of here anyway.

If this country needs me to fight then they will have to show me they appreciate it. The very next day the trip was made to the Council to present my demands. I am the king damn it, so they can treat me like one. Going into Sam's office we exchanged the usually greeting.

"Good to see you, Daniel. What brings you here today?" Sam asked.

"I came to see about arrangements for the fight. You guys said the organization of it should be left in your hands. That's fine, but there are a few small things I would like done." I said. A cart should be made to carry me to the square. It should be decorated with lots of bright colors. The handles should be long enough so at least six females can pull me up to the square. A gold-colored robe should be made for me to wear up to the fight area where another female will slide it off my shoulders, gently fold it, and then hand it to another female to put somewhere. An impressive show should send a message to the ambassadors from the other continent that we are equal to anything they can do. I will draw the plan for what the cart should look like. Those females should all wear colorful matching uniforms," I said.

I could imagine that it all should look very regal and with everything under the control of a strong handsome King. Just like some of the movies I had seen where the King comes home to his palace from a triumph of some kind to the cheers of spectators lining the way.

"Okay if that's what you really want. We will get started on it right away just as soon as you get a drawing to us on the cart," Sam said with a frown on his face.

That afternoon when I got back to the house, Cookie had a dessert, and drink ready. She really took good care of me, and that was much appreciated. If only she was a good-looking real girl it would be impossible not to make a pass at her in my lonely state of mind. I needed a queen to ride by my side in that cart. But a beautiful queen to show off to the public was nowhere to be found in this country.

I could picture in my mind one of those big black football heroes with the blond beauty on his arm making an entrance to a fancy nightclub. Must be very satisfying to have that on your arm, and also in the bed that night. His ego probably tell him it's because he is so good-looking that she adores him. But most of us know it is only his money that she really loves. Lust and love can both exist at the same time, but lust by itself is nothing more than rape without violence. One has a lust for fame and respect. The other has a lust for riches, and to show off his trophy wife that he had conquered as a hero.

Just then Cindy came into the kitchen, took a cup, poured some juice, and joined me at the table.

"What's wrong Daniel? You don't seem very happy today. Did you get a bad report from the science laboratory?" she asked.

"They can't locate my planet. They only stopped there because water was detected, and that looked like a good place to resupply. My order was for them to keep searching their records. I want to go home." I replied. "When they do a backtrack search on the earlier trip they should find my planet. I want a partner, but who here will have me with this different face? So I need to go home.

Walk with me down to the pond. I find peace and quiet there." I demanded.

We left the house, and as we started toward the path Cindy took my hand in hers as usual. We walked together along the path until getting to the bench to sit and look over the water. Determining to find out if there was anything more in our relationship besides just holding hands I came right out with it and asked.

"You always take my hand when we walk to the pond. Do you like me enough to take this relationship any further?" I asked.

"What do you mean by further?" I like you, but I need to know what you mean by further," she replied.

"I mean could you have a closer relationship? Would you consider me as a partner? How about sex, would you have sex with me?" I persisted.

"No! No! I said before that I liked you, but sex! Partners! No, that can't be. We are too different," Cindy moved away from me.

If there are no feelings toward me then why do you take my hand when we walk to this pond?" I asked.

"Because you are the King. When you told me to take your hand the first time we walked here I thought that was a command to take your hand every time we walked together to this pond, so that is why I do it."

"So that is the only reason? Why can't you think of me as just a regular guy and not a King? Then maybe we could take this a step further," I persisted.

"Again I must say no. that wouldn't work."

"But why? Why wouldn't it work?" I insisted on knowing.

"I'm sorry, but you are just too ugly. After all this time I still have trouble looking at you. Even with the nose reduction, your one nose is still way too big. And all that hair on your head and face. It goes all over the place and never looks clean. Even if you could lose the hair, and have a smooth head and face like us, that nose is still more than I could find attractive. I'm sorry Daniel, but you asked", Cindy explained.

I sat there in silence for several moments trying to absorb what she had said. The rage of such blunt rejection flooded over me. The urge to reach out and strangle her was too much to overcome so making a hasty retreat I hurried back up the path to the house without her.

The next few days I busied myself in workouts and conditioning routines, designing the cart and planning how the fight ceremony should go. Putting it all on the transfer machine it was sent to the Council office.

This had just been completed when the two guys from the audio lab showed up with a new translator. I kept the old one on me where ever I went. Some words were easy, but other words had such strange sounds with frequencies I couldn't even hear that conversation was extremely hard for me to understand, so this translator was always with me.

"Cindy has made some recordings of your conversation with her. Using those we could translate what you said, and get the meaning much improved. This new one should work much better for you. Cindy wanted to give you a surprise with this new one. I hope you find it much improved," one of them said.

Damn! I thought. And after walking out on her too. She is one thoughtful female. Somehow she will be rewarded for this. Fastening the new translator around my neck, and exchanging conversation with the two guys it was easy to see this one was indeed much improved over the old one. I thanked them for the new one, told them to turn in the cost to the Council, and escorted them to the door.

Thinking about the kindness of Cindy, and showing her I cared, a trip to the Council to tell Sam of my plan would be taken care of the next morning. Not only would she be riding in the cart with me, but there would be a special uniform for her as well.

As time came closer to the contest, a small contingent from the other continent came to work out the arrangements with our group. The first thing they did at the meeting was to present a petition to have me declared ineligible to compete. I did not go to the meeting but was apprised of the situation later. The petition stated that only certified residents are allowed to participate. Then the plans for how the celebration would progress became a point of contention. The spectacle of the cart and how it would be brought in would not be acceptable procedure. No agreement could be reached at this meeting. It was left that we would run the celebration the way we wanted when it was on our continent, and they could run it the way they wanted it when on their continent.

The day came for the contest. The opposition docked at our shore where our group met them with a show of courtesy and respect. By the time we had escorted them to their quarters, they seemed to feel that all would be well, and this would actually be very similar to past events.

When the morrow came and they were escorted to the square area, a huge crowd had already gathered to watch. The instructions given the spectators at the entrance were pretty much followed as written. The visitors were cheered and given a friendly welcome. So they were pleased with this reception.

The cart had been set up and ready to go. The females were dressed in bright golden-colored uniforms. Cindy was very surprised at being asked to ride in the cart by my side. Dressed in her bright golden-colored uniform with the red trim, and emblem on the front of it, she looked beautiful sitting beside me with my golden-colored robe with the matching red trim. Everybody had a big smile on their face knowing this would be a spectacular entrance.

Starting down the lane of spectators, there was a gasp of surprise at what was in front of them followed by loud cheering that followed us all the way to the square.

The females held the handles of the cart in one hand, and stepped in unison along the lane up to the area where they stopped in unison as well, and remained at attention. An aide stepped up to the side, opened the side door, stepped back, and placed a hand over his chest as a salute as I exited. Another male in turn stepped to the other side of the cart to offer a hand to help Cindy out. She was then escorted to a place of honor at the side of the square.

The opponents stood off to one side with a look on their faces that couldn't believe what they were seeing. Past contests had deliberately been kept as a simpler dignified affair befitting a king. This was going to be my last contest, so this would give them something to remember. I will be the

King for as long as I want, but no more fights for it.

The preliminary announcements on the rules out of the way, the introductions were next, the visitor first, and then me. My corner assistant stepped up, slid the robe off my shoulders, folded it neatly and handed it to his assistant, (just like the wrestler Gorgeous George did) who held it carefully as it was carried away. We faced each other in the middle of the square, gave the traditional greeting, and stepped back to our respective sides.

It was at this moment that even with my superior strength and speed the butterflies came to my stomach just like they did before a wrestling match started when I wrestled at the university. The signal to start rang out. We both cautiously moved to the center to face each other. Looking at this opponent, there was no sneer on his face like that one last year. The jaw was set in determination to do his best but it was obvious he was petrified at facing a larger, stronger opponent.

My heart went out to him. Knowing he would be badly beaten, and with what happened to the last guy, that kind of courage had to be respected. Letting him get in the first couple of hits it was easy to turn my head enough so they slid off with no effect. Putting in a couple of punches of my own, and holding back on the power made the bout look pretty good. The crowd seemed to be enjoying the action so this show was continued for a few moments. Getting warmed up, and feeling good, one of my punches went in too fast, and he went down pretty hard.

I stepped up closer to bend over and tell him to quit, but since he didn't look like it was time yet, I put out a hand and pulled him up to his feet. This had never been done before. The crowd roared at this show of sportsmanship. It was fun for me, so why not give them a little more of a show since this is the last one they would see, and this guy deserved some respect. I got a little careless, and got in too close. Those lower hands grabbed onto my torso and a few blows from his upper hands pounded on my face. Exchanging a few more blows, one landed pretty good on the side of my face, so pretended to stagger backward, and fell down. Hesitating just a moment, my opponent stepped up, held out his hand, and helped me to my feet. The crowd really roared at this show of sportsmanship. I loved it. He was returning the kindness. But it was time to end this farce. We had fooled the people into thinking they were seeing a good evenly matched fight. Deliberately going into a clinch I tried to tell him to stay down and quit the next time I threw a punch. Without my translator, it was hard to know if he understood.

I planted one the side of his head hard enough to put him down, and he had the good sense to stay there. Staying down for a sufficient amount of time I reached down to pull him to his feet. He indicated that he was done. To show respect the standard hands on the shoulder bit was offered and returned by him. The crowd gradually filed out satisfied they had seen some good competition, with very good sportsmanship to go with it, and pleased that the King badge would remain on this continent for another year.

When we were through cleaning up, and ready to go, I made a personal visit to the room of my opponent. There was such good sportsmanship on his part which was heads and shoulders above what last year's opponent showed that he was invited to stay the night at my house. The invitation was accepted. Instead of coming later, I ask him to ride with me right away back to my place. He even greeted Speedy with dignity and grace. We were hitting it off immediately, and with the translator now turned on the conversation was also easy.

By the time we got there, and introductions made of all the Council members, Cindy, and servants as well, it was time for dinner. Cookie had fixed a fabulous meal. The conversation was pretty much on general stuff, but before dinner was done it had switched to the King contests. Before it was over, we had all enthusiastically agreed that each continent should have its own King, and eliminate the contest. All except Cindy. She had repeated her earlier objections several times, but with this all in the distant past she gave up and went along with it. The annual celebration should be replaced with something else. Each continent will have their meeting and set the requirements for their own King. Jane showed our guests to their quarters, and the group broke up for the night.

The next morning saw the opponents pack up to head back to the other continent. It had been agreed that the idea of each continent having their own king would be presented to the respective councils. That was the last I saw of them or heard anything of their decision.

Our country seemed to be going along in slow motion. Nothing new was happening, or new ideas being proposed. With my trips through the countryside, there was a lot of beautiful areas just sitting with nothing being done. In the cities, there were also areas that was nothing but flat empty plots of land.

A few days were spent getting my ideas organized and on paper. A copy would be sent to each member of the Council before having a meeting with them. Cindy had been my trusted adviser so I wanted to run this by her first to see what she thought of the idea. So with a treat from Cookie before us, we sat at the table and considered the proposal.

"I want to organize a squadron of young people to clean up abandoned areas of land. Then plant things that would make that area more attractive. Areas would be fenced off for games. A walkway or path of some kind would be cleaned up and marked for self-guided tours. I don't see any of this being done, and I think it would bring pride as well as enjoyment to the people," I said to Cindy.

"Sounds like a nice idea, but we have not had any organized groups for generations. It is illegal to have any organization like that," she replied.

"But why is it against the law?" I asked. "This would be a great benefit to the country."

"I have told you before about the great wars in our history. They finally realized that all the death and destruction was not worth who would rule or control everything. To prevent another gradual progression to that end all organization for anything but education was made illegal. That was many generations ago, and it is the King contest that symbolizes the contest. And that is a celebration to commemorate the end of war. So no Daniel, it is a nice idea, but it would be illegal to organize a group to do it," Cindy explained.

"Then it's time that law was updated. Things change over time, and a country has to be ready to change with it. If it doesn't adapt to meet new challenges, then the destiny of any country is to fall behind. I will present this new idea to the Council for their approval tomorrow," and that ended the discussion.

So rather than get into a detailed discussion with the Council members, I decided to just do it myself. It took quite a few days traveling from city to city to locate, and convince a "volunteer" that it was there civic duty to organize the young people to do the job of cleaning up, restoring, and helping construct new features for that community. A small stipend and an "official" commendation would be awarded to the volunteer.

Thinking of past group organizations, and the effectiveness of organizing these young people it was the boy scout organization that I thought of, but it was the knowledge that this kind of activity would be accepted with reluctance, the idea of Hitler's Nazi Youth Core came to mind as the one way that would get results the quickest. It was very effective. It demanded loyalty and obedience. It established national pride, and it seemed like the most efficient way to get the job done.

The volunteers would be given the title of commander with a special uniform. The young people would also be given uniforms, and a patch that would signify what group they were a part of, and then a patch for the accomplishment of that particular group. The patch would be attached at the location on the uniform as determined by the commander.

When these projects were being started on the local level, I would select district officers who would take directions from me, and report back on any problems. In time as these various groups grew in size, and age, they could become very invaluable if needed to carry out an operation of some kind. As the King, they should feel honored to have me come to their area for a pep talk, and motivation for continued good work.

It was amazing how fast this idea progressed. Pride in their nation was developing faster than imagined, so it seemed like a good next step to introduce sports into the life of these kids. In considering what would be needed, the logical thing it seemed to me would be soccer. Not much needed for equipment, and any kid could kick a ball. Have the soccer balls manufactured; meet with all the district managers; hand out the rules; give a brief demonstration of the rules, and tell them the date to get started. Recruit the young ones in schools and other places where you can find them. A meeting was set for all district organizers to meet in the capital for information on continuing progress of the program.

It was amazing how much power a king has in ordering people to do things. Not wanting to waste this power, it seemed to me that the other continent should be getting their youth involved in some activities as well, and it might as well be soccer. Then in time, an international championship competition could be held. Going to the Council, they were instructed to call a meeting with the representatives of the other continent for the purpose of organizing young people to learn the game of soccer.

When the representatives met, the purpose of the local organization and the game had to be explained in detail on how it would benefit the communities as well as the kids. Community improvement projects would get done and the kids would have fun. There was some resistance on organizing a group for a particular purpose because of the past, but the wars had been several generations ago, and the memory of why it happened had faded over the years.

These people were physically weak with under developed bodies. The attitude of using only the brain, and not the body was foreign to me as I was raised to rely on both. But here even in jobs like farming or construction, machines did most of the work. What couldn't be done with machines didn't seem to get done at all. I was pleased that my superior strength got me such an advantage in the king contests, but to see these puny kids was a big disappointment. Hopefully the soccer games and training will lead to much stronger and better built bodies. Personally I don't give a damn about it, but I've got to do something while trapped here, so may as well try to enjoy what I can.

Enlisting Speedy to navigate our way around the country, we looked for places that needed cleanup work and would also be a good place to make into a soccer field.
Then the organizer in that area would be contacted to get to work on it. This was a very enjoyable activity as I got to see a lot of the continent. When an area was discovered that needed work, it was delegated to somebody, and I didn't have to do any of it myself. Before a year had gone by, this project was humming right along.

The big problem I hadn't counted on was the lack of interest in the lack of people wanting to do any physical work even when it would benefit their own community. The kids had no ambition. No incentive to get off their dead butts and join in the game of soccer. So to get some things done, and try to develop at least some social sense of pride and participation I called on my alliances for ideas. With no real understanding of public motivation, the problem was then presented to the council. They weren't any better since there was no training in their background either, and from what they were taught the idea of a group effort, or any idea of physical competition was met with resistance and negative arguments why this shouldn't even be considered.

But I didn't really give a damn what they thought. I was basically a prisoner here, trapped in a life I didn't want or understand. The only thing going for me was the power I had as the King, so I would use it if only for my own entertainment. If it was beneficial for them, fine, if not, who cares? I don't think they do, so why should I?

Taking a page from Hitler's playbook, if some kid didn't want to participate the others were taught how to shame him or her, and in some that just plain refused, a group trip to an isolated area would soon convince them to participate. This kind of Hitler persuasion was difficult to get into the attitude of many kids, but once they saw how effective it could be, then using a less-than-pleasant experience on a reluctant kid as a regular source of persuasion would become acceptable over time. So picking out some of the ones that seemed a little more of a rebel, some training was begun on how to be a leader with authority. Using the old phrase "a trip to the woodshed" carried out with severe enough persuasion and the threat of more visits to the "woodshed" and we built up a following group of participants.

In time, teams were established; games and tournaments were held; fans and supporters were building; empty lots were being cleared to make way for new recreation areas. I had little information on the other continent, but as long as this one was doing well it was a good feeling.

The commanders were instructed to have their group design a uniform that would distinguish them at meetings, and at games. The uniforms would be made and paid for with funds from the Council. This was the turning point between myself and the Council. The next Council meeting was a point of contention over who pays for what.

I AM THE ALIEN

"You have no authority to take on large expenses unless we okay it," Harry Said.

"You are mistaken on that Harry," I countered. "I am the King, and as such, the treasury will be controlled by me. I am the one allowing you to use the money to finance your projects and to pay for the businesses that you have approved. So far you have all done a pretty good job of running each department. You may as well know that I get a regular report from the treasurer. I didn't object when you all gave yourselves a raise in pay, and I didn't object when you increased spending for some pretty cushy perks. Just don't get too lavish in your ways."

It was obvious the entire Council was a little unnerved at the realization they were in for a power struggle over who would control the finances. I was here against my will and had this power only until there was a new king, so until then this place would operate as I wanted. I am the king, so I'll act like a king. Anything that happens here will not go down in history for anybody to read that means anything to me unless there is a return to Earth. I will only be a foot-note in the records of this planet.

With things getting a bit testier with the Council, and some negative feedback from some of the authorities I was glad I had taken charge of all the laser guns. As soon as I got back to the place that made them, they would get another order for some additional weapons.

It crossed my mind that my feelings for these people must be similar to what many of the white settlers thought about the Indians as they moved across the country to settle on new lands. Those Indians looked different, and lacked knowledge of the white man. So were deemed inferior, in the way of progress, so might as well get rid of them. Seemed to me they never stopped to think about how the Indian's knowledge of many things was superior to the whites.

There was no mistaking the disapproving glances and the utter contempt of somebody as inferior as they consider me to be that I should hold power over their lives. With each day the hatred for these natives with that scornful superiority grew within me. I had worked hard to benefit the "inferiors" and try to show the others that they were totally acceptable as individuals. But now even many of those I helped were beginning to show signs of rejection and contempt when looking at me.

       That got me wondering just what was the criteria for being labeled as inferior. I didn't ask as I didn't want to pry into a person's privacy, but so far nobody has told me what their condition was that made them inferior. I will find out what group makes that determination and get a written list to see what can be deleted or revised.

In my case here on this dinky little planet it was me that was so different, and me that was so inferior in technical knowledge, and it was very obvious to me that I was considered an inferior species. But I was so much superior in physical strength that I had no trouble thinking that I was the "top dog", and becoming the King solidified my thoughts on being the "top dog". There was no doubt in my mind that this planet could be shaped any way I wanted as long as the King position could be held, and controlled. At that moment in my mind I was superior all of this odd race in physical prowess, and probably in mental prowess as well if I had time to gain all the available knowledge they have had years to learn.

As the days went by the opinion of how things were being done was beginning to divide into two factions. Those that saw the improvement in the upkeep of the land with the work of the youth corp were pleased with that effort, but there was the minority that thought this would lead to a decline, and the reversal again to competitive factions that would pit one group against another, and continue deteriorating to the point of once again war breaking out as a way of settling disputes or superiority. Two of the Council members were also beginning to express doubts about the direction the country was going.

On earth, competitive activities only seemed to lead to advance better competition in sports, and better products in commercial competition. So when the subject of competition came up, my view was presented as the superior one. Any disagreement that came in written form was simply ignored. This didn't always go over well with some, but I wasn't going to be bothered with replying to such unimportant objections.

The time for another King competition was at hand again, and the other continent had sent their delegation here to work out the arrangements for the next match. I didn't realize they were here until the Council notified me that the match was being considered, and I should be available for consultation. I never waited for Speedy to get out to the car: just got in it, and sped over to the meeting by myself.

Emotions were sometimes difficult to see on the faces of this species, but stomping into that room there was no doubt there was concern on why I showed up so fast. So without wasting any time on formal greetings and niceties, I gave the short speech that had been on my mind for some time.

"There will be no king match this year, or any other year," I said. "It is long overdue that this charade of who shall be a King is ended. I am the King here and will be the King for as long as I want. If you want a leader on your continent then go ahead and choose one. You can do it with a fight if you want, or elect one by vote if that is your choice. But this world has only one King. We will not allow more than one. I am the King, and the only king. Anything else will not be tolerated. If you choose a new leader he shall not be called a King. I am the only King, and anybody you choose from your continent will be subservient to me."

I told the delegation from the other continent to go home; the discussion ended. I stood there frowning at them for a few moments until they got up from their chairs, and silently filed out.

I hadn't heard anything of having a King on each continent from them, and our inquiries were ignored. I was still seething at the treatment we received when in their country, so to hell with them. I was in charge here, and they would just have to accept that we were no longer the inferior species on this planet.

When they had cleared the building my own council members started in on me about abandoning the established procedure for determining a King, and how both continents should be involved in picking just one King, and that I had no right to do anything else but follow that protocol.

By now I had had enough of this place, and these people. They weren't people; they weren't humans. As far as I was concerned they were such a different species from me that I thought of them as such subhuman beings they might as well be labeled as monkeys or apes. I had to get out of here somehow, or do something but didn't know what.

So without a word, I turned my back to them and stomped out of the building. It seemed the thing to do was head to the science lab where the space exploration was located to see if anything had been found on that trip where they stopped for water, and I was captured. Nothing of that trip had been located as of that time. I sensed they actually did know, but for some reason were deliberately withholding that information from me. Feeling stymied by my situation, temper took over, and I blew my top. Demanding to know more of the location of Earth, and the logistics of getting there, I gave them an ultimatum on getting off their dead asses and laying out a plan to get me back home.

There was not much of a reason they should take on the task of getting me back though. After all, if they thought of me as an inferior species the way I thought of them, it would be a big saving, and a blessing if I just died right here. I wasn't ready to die, so if they saw no reason to get me out of here I would give them a reason. In the meantime, this place would dance to my tune, and not me to theirs.

Calming my rage some, I got in my vehicle, and headed home. When I pulled into my home place, speedy was waiting there with an agitated look. So I was informed that driving was not my job, but his, and he didn't want to lose his position. I was the King, but he felt comfortable enough to complain without fear of retaliation, but I'd had enough of the superior attitude of most of the inhabitants on this place. So I let him know that I was the King and the one in charge. When it was time for him to go, I would be the one deciding that, so quit worrying about the job until I tell you otherwise.

His eyes immediately looked away as though focusing on something that was way off to my left. I had noticed that look before on occasions when my guides or helpers were confronted by someone in a superior position was directing us on where to go, or explaining something. So after some consideration on using this to elevate my superiority and control over everything on this planet, I would issue a directive that when the king was present no eyes would be focused directly on him. In fact all eyes should be focused on the ground until an individual was addressed by the king, and a reply was expected. I knew this is the way it worked in some of our countries so it should work here as well. My anger at the do-nothing attitude and the contempt I saw in them for me as inferior led to a formal edict that when in the presents of the King all eyes would be cast downward to the ground and lifted only when spoken to directly by the King.

It made me recall on of my speaking times when the worst of the disrespect came through loud and clear. The insults that were said without fear of being overheard or stopped, insulted my entire being. It reached a climax when one of them placed a large replica of my original nose in the middle of his face right between both of his smaller noses. The guys around him were snickering and commenting on it while laughing. This was the last time a speech would be heard by me. I left there determined to somehow make them pay for that disrespect. I would have my revenge.

Since it would not be readily acceptable to all that was present in my circle of accomplices, they would all be free to disregard that law for our interactions until further notice. Our interactions had been on a rather informal, casual relationship with humor and occasionally a little "horseplay" mixed in. Much of the "horsing" around being instigated by me, and it had taken some time for them to get the idea of teasing each other could be fun. I enjoyed this interaction even though sometimes one of them would forget to keep it light-hearted, and it came through as a bit critical, which I did not appreciate.

I would have the council oversee that this new law was sent to all districts of the continent and properly distributed to the population. I hadn't thought out what the punishment might for this violation or punishment for a violation of any kind. That will have to be determined very soon. When I meet with the council that will be taken up for consideration, though I will make the final decision.

Cookie met me at the door to inform me that there were two visitors waiting in the dining room. Her idea of the protocol was to lead a visitor into the dining room to have a drink of her concoction that I thought of as tea.

Sitting there with drinks in hand were two of the inferiors I had met before. One of them was an employee of the place that reluctantly gave me my first opportunity to have a job. I couldn't remember his name, or even if he had told me a name. The other one was a stranger. Cookie made the introductions and then offered to get me a drink of whatever was in that concoction she enjoyed putting in front of anybody and everybody that set foot in her space.

By the time our visit was over, my understanding was that one by one all of the inferiors were being phased out to be replaced by the superiors. The fired inferiors had no place to go for complaints as they had no standing to ask for a meeting. They knew it would be a lost cause as they would not even be listened to by officials that were superior.

I instructed them to go back, list all the names of those working there that were classified as inferior; the percentage of the workforce listed as inferior, and get back to me with this information. Then we could consider a formal response to the company.

As long as they were willing to talk of those troubles I figured a few questions to them might give me some information on things I hadn't considered before and had no interest in.

"What happens when a crime is committed, or a major infraction of the laws has to be dealt with by the administration? What happens to the person." I asked. "I have heard nothing about punishment or incarceration. So what is done with them to keep repeat offenders from doing it again?"

They both stood there in silence apparently reluctant to say anything. Cookie had been nearby the whole time listening. She seemed to know a lot about everything that goes on here, and maybe being in this place where a lot of high-level meetings take place she has heard it all. Maybe they didn't mind her listening to the discussions, or maybe she had clearance to hear it all. Or maybe she was just plain nosy. It was no matter to me, but when she said she knew, I was going to be all ears to hear her opinion. So I asked her to explain the whole process on the procedure for dealing with a bad person.

Before starting, she placed a treat in a container for the two guys to take with them and escorted them to the door to leave. They seemed a little reluctant to go, but with Cookie's stern looks they departed to head back home.

"I haven't seen any places of incarceration or what we call jails on my planet. Bad people have to be confined or kept out of the populace. How is it done here to keep somebody from further disruption in society"? I asked.

"It is kept pretty quiet. We all know a disruptor is removed, but the Special Forces are not allowed to talk about it. I know because I was a member of the special force that dealt with the group handling female cases." Cookie spoke up.

"The criminal was taken to the northern border of our country which was surrounded by a high wall, and given the choice of instant death by chemical inhalant, or passing through the border gate to live on the northern side. They were not told what was on that North side, but sure death from the inhaled gas, or the chance to go free in the North didn't seem like a difficult choice. So of course, almost all chose to go through the gate. The wall was so high that it would be impossible to escape over the top, and the gate so heavy and strong it could not be opened unless you had keys to the big locks, and it could not be opened from the inside. This is why we don't need "jails" as there are no repeat offenders. And that is why there is no drain on public resources to incarcerate, feed, and care for anybody that has committed a crime."

She spoke as though she knew what it was all about, and that really made me start to wonder about her background and how she would know this.

That seemed like a very satisfactory way to deal with a criminal. They are out of society, and yet can then develop their own lives by living in the North. Sort of like sending the "criminals" to Australia as a solution; instead of putting them in jail and having to feed them as the British used to do many years ago. But that started me instantly wondering what it was like in the Northern region, so asked Cookie to tell me about it. If they were never seen again, they must be happy there, especially if nobody made a request to come home again, and another chance to prove they would no longer be a threat if granted the request to return to the South side of the wall. Cookie hesitated in silence for a few moments, let out a slow sigh and continued.

"Since you now consider yourself the permanent king, then you should know the whole story on it. Nobody has ever asked to return to the South as nobody has ever survived long enough to try to get back.

You know of the fungus that we have for food, and you know of how it grows and reproduces by scattering spores. When those spores land on a source of nutrition they settle down on it and absorb it into their system. You have seen how the lower appendage can expand until it explodes upward to scatter its spores into the air or to land on a food source.

We have the same thing here in the North except these fungus systems have modified significantly over time to be a much larger unit. They have also developed small hooks on each spore so that when it hits a food source the point of this hook penetrates and the hooks take hold to fasten itself securely. The spores then secrete a fluid that instantly starts to dissolve anything it is attached to so that the spore's cells can absorb the nutrition. That fluid is a very caustic acid. When that acid starts to dissolve the skin on a person it is very painful. It only takes part of a day before enough of the victim dies. The screams of pain can be heard for some distance. We never get used to hearing that, and it lingers in our minds for days. Some of us can stand it no longer so use the poison inhalant to kill ourselves to get away from doing this any longer. I managed to fake my death; make it back to home, and establish a new life."

"Is that why the fence is so high I asked.

"Yes, of course. It is not only high, but solid all the way to the top. We can't take a chance on any of those spores getting to this side where they could grow and reproduce. This isolation program started way back before my time. Any questions that were asked were immediately dismissed without any explanation. So I have no information on how the wall came to be constructed or why I was picked to be one of the members on the removal squad", Cookie's voice faded into silence, and she looked as though she was physically fading as well.

"Okay, that tells me quite a bit of the process, so tell me something of who are the primary targets of this barbaric elimination. Is it just the severe bad people, or what is their crime?" I asked.

"It is just the ones that are labeled inferior that get sentenced for extermination when they commit a crime. In most cases, it is only just some small infraction that gets the attention of the enforcers. It would have to be something very serious for the upper-level person to get arrested. In fact, I know of only four of them being taken to the wall while I was in the force that had to do the elimination. It was never a fair choice. The inferior was always found guilty and had to pay the supreme price. I always did think this was a way to get rid of all inferiors by taking them out a few at a time instead of rounding them up in mass for extinction. If we inferiors had not been needed to do most of the work I'm sure it would have been much different for us," Cookie shrugged her tired shoulders. "But what can we do about it? We are helpless with no power to change it."

"So how did you get this job, and why aren't you still working there now?" I wanted to know.

"I don't know how or why I was chosen, but the job came with lots of extra privileges so at the time I was happy to be selected. I soon regretted it when I then found out you can't quit that job, were sworn to secrecy under penalty of death, and other conditions as well" cookie explained.

"May I go see this wall, and see what is on the other side," I asked.

"No, no, I don't think you should even consider it. It is too high to see over the top, and solid all the way through, so you can't see anything that way. No one would let you in through the gate because you would not be permitted out once the gate was closed behind you," she explained.

"Okay, I understand the danger of these things but from what I've seen of the fungus around here, the only time that lower appendage explodes the thing upward is not very high, and it only moves a short distance when the wind is blowing. So all we have to do is stay a short distance from them." and cookie started to talk before I was done with any reasoning.

"But it's not the same behind the wall," she continued. The fungi on the North side of the wall grow a lot higher and when that appendage fires off it can go higher and further. It doesn't have to wait for it to reach full capacity. It can trigger an explosion when it wants. It seems to sense when a food source is near. It is not totally at the mercy of which way the wind goes. That appendage can act as a bit of a rudder to guide it in a slightly different path. It can also fire off its' explosive leg again and again. It does get a little weaker each time, and can't go as far, but that's still enough to get it to a food source pretty quickly. Once it hits you or any part of you, those hooks will grab onto your skin, and then no matter what, you will be absorbed. So no, even though you are the King I will not guide you to that wall. I will not wait to hear you scream from the pain of absorption" and she turned to get me a refill on her "tea".

I wanted to learn a lot more about that place but wondered how much more information I could pry out of her. Decided to try for some personal information then to see what would come through on the wall. She offered some further information that there was a deliberate large border along the wall in case any fungus somehow managed to get out and start to establish a new colony of fungus. So periodically a crew of special workers would inspect the borderland and eliminate any new growth.

"How did you manage to get back to a regular spot in civilization then? You must have had some kind of help", I suggested. "I want to know how you did it. If it will help me be a better ruler of this society, a knowledge of what you all have to go through, and how it can be changed for the better would be most helpful."

"I told you how I pretended to breathe the gas, and then stumble down into the ravine to perish. I didn't die of course, and just kept going until out of site from the others, then hiding by laying down in the brush until they had left. I worked my way down further into the ravine and then headed South along the bottom so as not to be seen.

The second day making my way through the brush and rocks I stopped for a rest. It was very tiring going through that stuff so I needed a rest. With no food that whole time it didn't take long for my body to shut down. I woke up with a male poking at me with a stick. He offered to take me to his house to eat and have a cool drink. Being in no condition to resist, I jumped at the opportunity for some nourishment and rest. His wife was quick to notice my condition and also offered food and drink.

When they both left the room while I was eating it didn't seem odd to me at the time. But when I had finished they demanded that I go to a room to lay down and rest. Being very tired from my journey it would be a great relief to lie down for some rest. The shoes were kicked off to prevent dirt from soiling on the top cover of the bed.

My nap lasted a little longer than intended but was much needed. I awoke to a funny burning sensation in my feet, and on getting up it hurt really bad as I tried to stand. My shoes were nowhere to be seen, but in a moment the door opened and the man came in carrying my shoes. He told me my feet would hurt if I tried to walk very fast, and would be extremely painful if running on them or tried to get away. He planned on making me a prisoner for a while to help with picking the crop of berries that he sold.

I learned that the potion they applied to the bottom of my feet was from a plant that was very acidic. Any wear and tear on my feet would result in skin tearing away and bare flesh would be too painful to endure. So if I behaved myself for several days that when the berry season was done then I could leave. In the meantime, my job was to help in the house and also help pick the berries.

Even with foot covers, wearing the shoes was pretty painful. Going out to pick berries was the worst, as the rough ground felt like hammers hitting the bottom of my feet. Mornings were the worst as they made me be still while a new layer of that acid was applied. I tried to avoid some of it, but his impatience at my reaction when the new stuff was applied brought a sharp blow to my face. I tried to grit my teeth and get ready for it, but most days brought the hard face slap anyway.

Several days went by before I noticed another plant growing in the same area as the berries. I remembered much of my chemistry class at school, and this was a plant that had a sticky whitish solution that oozed out of the stalk when squeezed or stepped on. It makes a mess on your shoes, and is difficult to clean off. I made a point to scuff my shoes on that plant before going in with the berries. Taking my shoes off I made a point of leaving them in a slightly out-of-the-way spot. Then before going to bed, as much as I could get was scraped off the shoes I had hidden, then at bedtime applied the stuff to the bottom of my feet, and carefully rubbed all around the soles. After a few days of that treatment, the morning ritual of acid application was becoming ineffective. I pretended to hurt now, and was not hit very often. I hoped the berry season would end soon, but kept up my own late-night treatment of the sticky stuff. The "pain act" out in the field was emphasized. The change in color of my feet was credited to what the acid was doing to the skin, and they bought that idea without any skepticism, though I knew it was from the sticky stuff that I was applying each night. It didn't take too many more days, and my mind went to escape plans.

During time out in the field, it was easy to look around for the best path to get out of there. I thought about putting a bunch of that acid in there bed somehow before I left, but did not want to get caught, so canceled that idea. Just wait until we all went to bed and then sneak out the door. So a small bag of food was stored beside my bed, and that night I took it and everything that was mine, and left.

There was no way for me to determine where the bigger cities were, or any non-safe routes so just started out across the ravine, up the other side and kept going. Anywhere but my old home place should be safe as it had been so long that nobody outside that area would recognize me. As long as my food lasted the cities were avoided. I just kept walking. Finally, there was a large nice looking house in the distance. It seemed pretty isolated out in the country all by itself. A short distance away there was a small lake and a bench along one side. That bench looked so inviting, and I was pretty tired so I headed for it, set down, then lay down to take a short nap.

A little later I was woken from my sleep to look up at an old lady standing in front of me. She was too old for me to be afraid. She sat down beside me and before long we were chatting like old friends. She explained what this place was all about. How the King would live here if he ever won the fight. She was the cook but hadn't had much to do lately. The Council was going to retire her from duties but would wait until there was a chance there would be a new king coming before letting her go.

I didn't tell her my whole story and fabricated most of how I got there. There was empathy in her voice as she invited me to stay the night, and have food and drink with her. Since there was really no where for me to go, and she seemed to need some company so I stayed. She eventually became too infirm to manage the job. When the Council decided to replace her it was her high recommendation that since I was familiar with the place the job be given to me. That is why I was there when you came."

Even though I would have liked to learn a lot more on the history and how genetics progressed on this particular type of fungus apparently as far as Cookie was concerned the case was closed. After she had topped off my cup I motioned for her to sit in the chair across from me as I wanted to bounce my idea of commanding respect from the populace at large. I had gained a new respect for this lady and realized that her knowledge and experience of this world was far more experienced than I had known, or even imagined.

The very next day I went to the council building to get them started on the new rule of not looking at the King. When the four of them were in the room it was explained how this rule would change the way this King would be treated and also create a more respectful attitude toward the inferiors and how they were treated.

They each had to take a turn with a story on their experience with discrimination. Things had been this way for so long that it seemed that a change was inconceivable. It has always been this way and it always will be. That has to be your station in life, and it can't be changed, for that is the way you are born seemed to be the unanimous position.

To me, this was not acceptable. So I said that this attitude was going to change whether it had always been that way or not. What I wanted done, and how they were going to help do it was explained in a matter-of-fact way. They were to start getting the information out to the stations around the country tomorrow. Anyone that did not follow these new rules would be arrested and suffer the penalty of disobedience. They were given a deadline of two days to decide on the penalty, and I would be back in two days to finalize it.

Leaving the room to go back home, Sam followed me out of the building and followed me toward the car. Before I could get in he took a stand, and it was obvious the intention was to ask about something. But there was no question.

"We can't do that," he spoke up. "We can't dictate to the superiors how to behave, or act when seeing you or anybody else. They are the ones in control of things, not you. We control the Council because what we do is kept low key, and any information that goes out is under the guise of sounding like it comes from the old council of superiors. You are only a king in name. This is an honorary position and has never been intended to have any authoritative power."

"Well Sam that has all changed. Now I not only have power, I am the power on this continent. What I say goes, and you will do your best to carry out any orders you are given," I replied.

"But you have no authority to give me or any others direct orders without the council's approval," Sam insisted.

"If you want to continue being a council member then you better do as I tell you," was my reply as the vehicle door slammed shut behind me to go home.

Pondering my options on the way there would have to be a force of some kind behind me to guarantee there would be hell to pay if any resistance was shown to any of the orders. I still had the guns from the earlier episode, but maybe now a smaller, easier-to-hide gun could come in handy. I would make the drive to that factory visited earlier where the guns were stored. If a small gun was available I'd take one of those. At this time I wasn't worried, but better to be prepared just in case.

Getting home there was Speedy standing there in front of the garage waiting. I had to smile at this dedication to the job, so vowed to give him priority to transport me to the places that called for my presence. He knew the roads of this land and just might come in handy at a future time. Best to stay on a friendly basis to keep him as an ally, and getting information that might be useful I would defer on some things. So smiled at him with a greeting while expressing regret at going without him, and indicated it would not happen again unless it was an emergency.

I was happy to have the next couple of days free from any connection with the Council, or any of the troubles that the group from the South was having with their employment. I didn't even want to think about it. So put it all out of my mind, and decided to head for the lake. There a person could sit in peace and quiet to get away from it all.

I headed to the door with my thoughts already bouncing in my head when thoughts turned to Cindy. I always seemed to have her with me when I took that walk to the lake. So I stopped and called her name. With the door held open with one hand and waiting, I called her again, but a little louder. Waiting there a little longer I began to wonder if some feelings were beginning to develop for her, or was my hatred to this uppity culture changing my feelings to distaste for this place with only a feeling of possession of this helpful person that not only does my bidding, but helps carry out any of the endeavors I think we should be doing. Oh well, no matter whether I use her or abuse her, she is here to be my helper and will do as I command. Granted that she has never made fun of me, or put me down. I still have hate for what this culture has put me through, and she is one of them. And yet she has been kind to me, and helped in ways she was not obligated to. I readily had to admit to myself that I didn't understand how I could hate, and yet love somebody at the same time.

She came through the hallway and walked out the door behind me. My feeling either way melted to nothing, and an old habit took over as reaching for her hand was automatic. Realizing what had just happened, my grip went limp to retrieve it to my side, but there was no release, as that dual thumb grip of hers held firm. That was always a good feeling to have her hand firmly in mind, so told myself to knock off the confused thinking, and get a grip not only on my confusion but on her as well. We continued down the familiar path to the bench by the lake.
We had been sitting there side by side in silence for some time before I decided to tell her of my plans about showing respect for the King.

She listened quietly, but it was easy to see that resistance was increasing to any plan of punishment. When my explanation was done, I asked what she thought of the plan.

"Why do you need a show of respect?" she asked. "What you do will get you respect or disdain depending on whether you are successful or not. If someone gets punished for simply looking at you, that would get you no respect. In fact, it would do the opposite, and any talk with other citizens about you would only be negative. So I believe the punishment should only be used when a person deliberately does wrong."

I explained that she was probably right on just looking, but mocking or in some way making fun of my looks would definitely not be tolerated. That sort of conduct would be subject to punishment. That kind of behavior had been increasing lately, and the disrespect was going to stop. I had been the subject of racial intolerance for too long. In my mind, they would deserve any punishment meted out.

I reached over to take her hand in mine as I thanked her for supporting, and helping me all this time with no complaining or objecting. I realized all that was ever offered from her was a hand. I wanted more than just a warm hand. Give me someone to hold, someone that would hold me back. I tried not to think of the women on Earth; what it felt like to hold one in my arms; the warmth of a cheek against mine; the imagination of laying in bed together with naked bodies pressed together. I wanted more, but there was nothing here. Anger was building up at my plight in this uncaring, unfeeling place. I know it's not their fault, but damn it, I have given them my all. I fought the king fight for them. I brought them the respect they wanted in this world, so why shouldn't I be entitled to have some for myself. So to hell with them all. I will have what I want with what means I need to use to get it.

Getting up to go back to the house, her hand was turned loose, and I refused to hold it on the way back up the path.

All the insults and jeers began to build in my mind. The laughter at my appearance, especially my nose, even my hands, and fingers took verbal abuse. The phrase "one thumb dumb" had been sent my way when introduced at speeches. Nobody had the guts to say it to my face, but hiding in a crowd those jerks seemed to feel safe yelling from there behind some cover of who was in front of them. I didn't understand exactly what was said, but the translator I always wore converted it enough that I could put most of it together to figure out what was being said. Their thumb on each side of their hand looked as silly to me as my one did to them. The time will come when they will regret it.

Getting back to the house I needed time to think through some things so excused myself, and went into the bedroom to sit by myself to run various scenarios through my head. Every scenario ended with having the strength of some kind of force to back it up. This led me to thoughts of the guns at the place I had visited earlier. A trip there would tell me how they work as the one visit did not tell me much. At that time the need to see what would happen, and how these so-called "people" would treat someone that was of a totally different species rattled around in my head that it had to be experienced by them. I was going to give them one more chance to show some respect without demanding it.

Speedy was summoned to drive into a busy part of this area; find a spot to park; wait for me there until I returned. I did tell him of my plan to go among the inhabitants to see what their reaction would be. There would be no disguise or covering worn. So with only the same clothing they wore, and with head and hands visible to see, I walked along the street with head held high and tried to look confident in my walk.

There were plenty of stares at my face and head, but comments were quietly whispered to each other when they were behind me. The next test was to go into a social place where there was food and drinks available. I don't know exactly what it was called, but on Earth, we called it a restaurant. Taking a seat I asked for a drink that seemed to be popular.

Lots of stares and whispers here as well, but it didn't take long for quiet insults could be heard with most directed at my nose, facial hair, and ears. I just pretended like these insults went on death ears, but when several came to my table to spout off on someone that offered no harm to them it finally began to get to me. With some insults of my own directed at them doing the best I could with that damn translator, it didn't take long to get physical. I was much bigger and stronger so conjured up my version of an old-fashioned bar fight from earlier fights I had seen in the good old western movies and waded into them with fists flailing.

A lot of damage had been inflicted on these tormentors, but higher numbers prevail until they were getting the best of me. It was about then that Speedy came in. With a lot of loud demands on his part, they all backed away but waiting to see if I was done or going to start it up again. Speedy admonished me for what was happening; then turned to loudly give the others a lecture on how to treat an outsider. I didn't see any resentment on even one face or any expression of concern over my treatment. And I sure didn't resent any blows that landed on any of those "two noses" faces.

I was grinning from ear to ear on seeing this "dressing down" of his fellow inhabitants. I never could think of them as "real people" and it was getting easier with time to think about them as non-people. I don't know what to call them, but a dislike, even hatred, has been working its way into my thinking on what to call them. So they will be referred to using one of the most hated slang words I know from planet Earth. "Gooks", they will be called "Gooks" from now on. They will have no knowledge of what that word is about, so I can use it to satisfy my hatred, and at the same time have something to name them as I write about what I observe and any interaction that goes on.

Contrary to what I have been told of how peaceful this country is now, from their story of the earlier wars, this was not always so. Also contrary to what I was led to believe, discrimination has been a big part of this country. Even reminds me of the Hitler regime in Germany. Genetics is a big part of this place. It may not be predicated on looks with preference of white skin and blond hair, but it sure has been employed in the discrimination of intelligence or to prevent certain defections in bodily disease or functions that happen with genes that carry that trait.

Anyhow, Speedy got me back to our car, and we got the hell away from there as fast as we could to get to the safety of home. I was ready to sit back and relax after the "dust-up" with the natives, and have a nice big drink of Cookie's tea stuff. So as soon as we got home, I motioned for speedy to join me so we could both have some downtime. He followed me inside to the dining room, took a seat, and started looking for Cookie to show up with the container or juice in hand.

Cookie wasn't too far behind and dutifully poured our drinks. Taking another look at my out-of-the-ordinary neat appearance she had to ask.

"What happened to you?"

"Got into a bit of a tussle. Thank God Speedy showed up when he did. Thank God he got us out of there in a hurry," was my answer. That was my sincere serious thought at the moment.

Cookie got her own serious look so I could see it coming. I'd been through a bit of discussion with Cindy, and realized I'd made a mistake just mentioning the word God, and knew Cookie would have to ask.

"Who is this God you are thanking for your rescue? I thought you would be thanking just Speedy for getting you out of trouble," she said.

"Nobody important. Just a matter of speech," I tried to avoid it.

"I heard you talking with Cindy about it, but couldn't understand what it was all about. Who is this person God? I've never heard of him before," she persisted.

"Just an imaginary being that helps people in need. Today I was in need so he sent Speedy to help," and made my tone of voice sound as though that was my final comment on it as I was at a loss to explain to an alien what a belief in God was all about. Maybe sometime in the future I can probably do a better job when it becomes more clear in my mind just what a God is.

I could see this wasn't going to satisfy her, so got out of it by explaining that we would go into it at some other time as I thought to myself, "God, get me off of this planet. I want to go home."

In a state of indecision on what I wanted out of my life here, a decision would have to be made. I must either go all in, or totally abandon any ambition to expand my control and total authority over these people. Man I hate referring to them as people, but don't like the designation "gooks" either, so for the rest of the recording of my experience on this planet, the term I'll use is "people".

I have another day before the two guys get back with their report on their place of employment, so will do some heavy thinking on what I want to do. In the meantime, I'll bounce it off of Speedy, Blackie, Cookie, and Cindy. Then maybe see if the Council will back me on what I decide.

When the two guys got back the next day (wish I could remember names, but I can't even pronounce them even if I could remember) they were ready with the numbers. Forty-two percent of the workers were categorized as inferior, thirty-one percent were not married, and so were not classified either inferior or standard. The rest we could not get enough information to come to a conclusion on how they should be classified. All the supervisors and managers were not listed in a class, as it was common knowledge that all of them would be standard, and hence exempt from a label. There was not one inferior in a position of authority.

In the discussion that followed this report, it was determined that there was much unrest among the workers at how unfair it was. Some workers were doing the work of managers and supervisors who got credit for it, while the worker had to just put up with the way it was, and accept their position in life.

The next trip to the Council building all the members were present, and waiting for any information on what I wanted done on this "respect" idea. Turned out that nothing had been done on getting any information out to the public on how to show respect to the King. My conclusion was that they were stalling to see if it would just go away by ignoring my directive to let time fade this idea away so nobody would have to be bothered with it.

I let this slip by for the time being and decided to bring up the unrest over the unfair working situation in businesses and factories. None of the Council members could quote, or even find rules or regulations regarding inferiors in the work place, nor any qualifications for managers. So it must be up to the owners, or group of owners to decide for themselves who will be doing what.

I gave them a lecture on how unfair this system of employment was to the inferiors. This has to change. They were charged with carrying out the task of writing a set of rules and regulations; getting it out to all places of employment and posting in public places. But it has to be approved by me before it goes public. Up until then, I hadn't given them any time for questions, so then asked if any clarification was needed, or if there were any questions on what was expected of them.

I had expected a stony silence, but I guess they were getting comfortable enough with my easygoing approach that this idea triggered a bunch of questions on how this will be enforced, and what about any exceptions for education, training, or experience.

I spent some time telling them that was up to them to put some thought into it so that the process seemed fair to everybody. Past classifications would no longer be considered. Education, training, and fitness for the job would be the only criteria. If an apprenticeship was offered then there would be no criteria considered. The only consideration would be if a candidate could successfully complete the apprenticeship.

Before leaving the room, once again Sam was an aggravating antagonist by claiming I had no right to order this, and that the public had no authority to carry out any of these actions. I had no intentions of getting into an argument or modifying my orders, so laid it on the line for him.

"I am the King, therefore you will carry out any orders I give you, or face the consequences of your disobedience. Do I make myself clear," was what my reply was to that little bit of insurrection?

With that final exchange, Sam was left to his own thoughts as I headed to my transportation to head home. I sure could use a good shot of bourbon about then, as speedy flipped the switch on the power to drive us away from there.

I needed some feedback on the exchange with the council, so Speedy was the target of some of my questions. At first, he was a little reluctant to answer so I came right out and ask him if in a confrontation would he side with the council, or if would he back me.

After some hemming and hawing around the admittance that he would back me was a welcome relief, because if Speedy would not be firm support then I would have to abandon my efforts to do anything of my own on this planet. And I still didn't know if Cindy or Cookie would back me. The decision was made to gradually increase my power until nobody would cross me without fear of losing their own standing, or maybe even their life. Though in the time I've been here it seems to me there is not a lot of value placed on life.

In case it became necessary that I had to protect myself, a backup of force seemed of primary importance. So it was going to be a trip to the armaments place to see what was available. Speedy would know how to get there. This time he could accompany me inside. He may be needed sometime in the future to be my backup force in case of a problem out of control. So he better know how the guns worked, and how to take care of them.

When we got to the factory, the manager recognized me from the previous visit. That wasn't hard to do of course when you have a face like mine, and probably the largest body he has ever seen. With the introduction of Speedy, and a short explanation of why we were here it was obvious that there was a reluctance to take us in to the room where the armaments were stored. It was a reminder of who was the king, and my authority over all activities in this country that got instant access to the room.

He knew that I still had several of the long guns that had been used once before, and quickly asked what we needed to know now. I won't go into a lot of the discussion here with him. So I will tell you that we came away with several handguns and a supply of the energy pellets that was the ammunition. These pellets provided the energy to launch, and once in contact with the body to deliver a massive death charge. I didn't understand how it worked, but knowing it would wipe out an enemy was understanding enough for me.

On the way back, Speedy tried to give me an explanation on the physics of these pellets. I didn't understand most of it but pretended to, and thanked him for the explanation.

The subject was changed to make sure he knew that his job was to also back me no matter what, as well as drive. I explained in no uncertain terms that this country needed a strong, tough, permanent leader, and since I was the king, that leader was going to be me. Any objections would be corrected immediately even if the objector had to be eliminated. Seeing Speediee's expression, Cookie's experience came to mind, so hastily added that what I meant by eliminated as an objector would simply be relocated to an area where they could live under the old rules. If they came to the understanding that my rules were better than they could come back in our society. This seemed to satisfy him for the moment that there would not be any violence for an objector.

Nobody on this planet had one of these weapons but a small group of law enforcement officers. In the next few days, Speedy and myself would be getting in some practice on using one of them.

The next time I got to the Council it became apparent immediately that they were in no hurry to carry out their orders. My anger at their dragging their feet welled to the surface, and I cussed them out the best I could trying to make it plain what I was saying with that damn translator hanging around my neck. This didn't have the desired effect I expected as they looked at the floor, or up at the ceiling as though ignoring me. Turning on my heel, and stomping out through the door, Sam followed me out and to the car where Speedy was waiting.

Then using some stern language of his own, proceeded to tell me what they would or would not do. This looked like it needed some individual attention, so I invited him to come to see me tomorrow. I would have Speedy pick him up to bring him home for some discussion on it. I was sure he could be convinced that my proposal was the best way to get this place under control and benefit for all.

When he got there the next day, instead of going in for one of Cookie's treats I directed him to go with me down the path as there was a nice sight for him to see. Motioning him to sit on the bench, I took a spot beside him and pointed out how beautiful it was to sit here and just look out over the lake. After some more small talk on the scenery, we turned to the discussion on yesterday. There was no persuading him on seeing it my way. When he came right out and said the public would not take those kinds of orders from an illegitimate king; especially one with such different features, and only the one ugly nose, that did it. My blood boiled over and If I had brought the gun with me, it would have been used right then and there. It was at that moment I remembered how fragile their brain and skull was. With my superior size and strength, I had knocked out my opponents in the fights with one good punch.

As we rose to go I turned to face him with an angry stare. When he met my gaze with a look that conveyed no fear of me that was the final blow to my authority. With a mighty round-house swing my fist connected with the side of his face which snapped his head sideways so violently that I knew he was dead by the time he hit the ground. It took a few minutes to decide what the course of action should be to justify his death. I would have liked to just get rid of the corpse by dumping it in the lake or a ravine, but decided to move him a short distance up the trail and claim he stumbled and fell: probably hit his head, and that's what killed him.

Speedy helped move the body to take back to the Council. He backed me up on the "accident" with the Council members, so I was relieved, and pleased that he could be counted on in case of a future "accident." I noticed some of the council members looked Sam over pretty carefully, but I think they were satisfied with the story since there was no visible signs of injury or marks that would make them suspect otherwise.

The council seemed reluctant to get started on the orders to inform all businesses and factories of the new regulations on hiring; discrimination on jobs assigned, and promotion to a management position. So it was time to take another step in the process.

A written command was sent out to all units that might be involved to select one of their peers to be a delegate to a national meeting. I would personally address the group on matters of great concern. But I needed a speaker that had a good command of the language with a good personality to deliver the message so that there would be no misunderstanding with my stumbling over how to pronounce every word. Even using the translator I had trouble being completely clear on what I was trying to say.

It was time to get Cindy back into the planning, so instructed her to go to the larger cities, and areas of manufacture to set up a schedule for the speaker. I would help the speaker craft a speech that would instill a feeling of responsibility for his fellow workers, and a fierce patriotism to defend the group. Since Speedy had the best understanding of the geography, and roads to get to these places he would drive her to these destinations. He would also be assigned that same job when the speeches started, except he would be taking the speaker to those locations.

Cindy wasn't very enthusiastic about doing this, but I insisted and also insisted she was to give me some feedback on how the locals were accepting the idea of a speaker coming to their area. While this was getting organized I would go to the Council and work on them to get behind me on this, and take an active part in developments.

When I got to the Council, the feeling there was that I was overstepping my authority. There was also some suspicion on Sam's death, but nobody had the guts to ask specific questions, so let this slide by as no problem. Bert was the braver one, and started the discussion on who really was in control of things. The public needed reinforcement on who was running the country. We had not had the king competition for two cycles now, and attitudes were beginning to form that things were just drifting along with not much control. Bert thought we should start the competition again, and people could focus on that. It would be something they could see, and take part in, and have pride in their local event.

I could tell that confidence in me as the king was slipping. The glances I got in public was one that was not in awe of a powerful king, but a look of indifference, or maybe even of derision that this monstrosity of an outsider should carry any authority or control over their lives. But I had brought the King position back to this side of the world through my own efforts, and the least they could do is show some respect for that. Any further progress gained while in office has also been because of my efforts. My mind is beginning to see only a bunch of ungrateful people that have no interest in promoting fairness or even harmony among themselves. You label them inferiors and misfits. You take away their individual rights to find a partner to love and bear offspring.

So why should I care about a populace that isn't even human? I believe I will approve of the King competition. I will even enter the local events to go through the elimination rounds. At each level, my superiority will be reinforced, and when I am the finalist they will find that dealing with a king that has complete dominance will be at my discretion.

So the Council was instructed to start organizing the competition. I would make the trip back to the home of my "parents" and enter the fight there again. The thought of seeing them again had never been pleasant, or even a desire to restore a relationship. They had done their duty to house me, and get me started on the path to survive on this planet. It was not loved for me, but a duty to the country's Council to succeed or risk the consequences of failure.

As the next few days went by, however, I found myself thinking about what they had done for me, and the memories of exchanging ideas. Ideas that were completely foreign to both of us. By the time we had parted ways maybe there was some feelings of love toward each other developing there. It was hard to tell, but I recognized those feelings of fondness and caring actually were there from my side of it. So it would be good to see them again; to discuss what my life was like now; to see the same support in the fight that was shown the first time I entered the competition. Maybe even take Cindy along to see her parents and some of the old friends she knew.

It would take some time for all these activities to get going. Time to line up the venue for the speeches; time to find the speaker and train him to say what we want; time to get the fight competition arranged; and time to get ready for all of this. And there hasn't even been enough time for Speedy to get in any practice with the guns.

I needed some time to get all this activity straight in my thinking, so this next day I walked down the path to the bench by the lake where there was peace and quiet that should help in thinking it through. I would try to control my anger at the insults about my looks. I would try to control my anger about comments on my trouble trying to pronounce their words correctly. Physically my vocal structure could not make the same sounds. The snickering I heard when I tried to communicate but the correct sound couldn't be produced was beginning to weigh on me. I was having difficulty trying to control my anger at comments on being such an inferior race that I should not be permitted to even be on this planet. I don't want to be on this planet, and the only way I know how to get off of it is to make such a problem that they would be happy to get rid of me. My conclusion so far is to try to force the aerospace program to retrace the previous route and let me off on Earth.

I have been conflicted on what to make of this place and my position in it. There have been moments of sympathy and support for me, and there have been moments of extreme bitterness and loathing of my being here. There has never been any show of wanting me on this planet. The support shown has only been that since I am here I should be tolerated. I am made to feel welcome only if and when something is done that improves their lot in some way. The next few days will shape how I cast my lot on this planet.

My last attempt to find some love and a person to love was a total failure. I went to the clinic that had operated on my nose to make it smaller. It was some smaller, but still larger than anyone else on this planet, and only one nose so the snide remarks still stung whenever I heard them. With a diagram drawn out, and a detailed explanation of what I wanted them to do in hand, I made my plea for them to construct a replica of the earth woman's reproductive system. Skip the whole baby part; just the sex part.

If that could be done, then finding a female that would be willing to undergo the surgery for the benefit of her King should not be too difficult. Looks didn't matter as these females' looks were so similar it was often difficult for me to tell them apart.

The first doctor I talked to found it an incredible idea that they should cut and modify one of their females just to make it possible for an alien to have sex with one of their females. Excusing himself immediately, it didn't take long to return with three other doctors that I presumed to be surgeons. One of them I recognized as the surgeon that worked on my nose. They insisted I explain what I wanted them to do. After going through it all again, it was explained to me that it could not be done. That much alteration of their women would be fatal. They wouldn't even consider it. I shouldn't even ask it. It wouldn't make any difference whether a King or not, they would not try it.

I didn't know of course if they were lying to me because of the thought of having a hideous-looking outsider with one of their women, or really couldn't overcome the physical differences in our two distinct bodies. I suspected the latter, but nothing could be done about that, so thanked them for the consideration, and got out of there.

On the way home my mind spun round and round on my life here. Cindy had seemed to be my friend but had no love for me. It seemed to me she only thought of this as a "duty" expected of her and society. Everywhere I went I felt revulsion of my looks, and only a duty to listen when I spoke. Everything I experienced at my home here from these workers seem to be done as a condition of employment, and not from any loyalty or love for me.

I had come to accept this treatment of an alien outsider as the norm, but I so much missed having a warm loving person to hold. The women I had known on Earth were missed more than I would admit to anybody here, but my heart, indeed my whole body ached to touch and hold them in my arms one more time. So what is there for me now on this planet?

The arrival of the representatives from the factory that was accused of using extreme amounts of discrimination brought things back to some serious thoughts on how to handle that problem. Their numbers on how many that were being caught in the web of discrimination was right on target with their first estimate. I had already decided that the best way to set things right was to use the same tactics as had been used throughout most of history by the depressed and mistreated masses. But we will try to make the change faster, and with less total violence. The strategy would be used instead of violence when possible.

"The strategy would be very simple I said. "Get back to the factory; explain the tactic to the workers; select the ones reliable enough to carry it out; let them know they would have my backing; pick a time to do it, and get it done. Before any action was taken, make sure you had picked a replacement for every manager and officer in the administration so the operation could continue running with a minimum of problems.

Start by getting the administration into a room for a meeting. Then have your chosen group of supporters file in and take positions around the room behind the company bosses. Explain to the bosses why this meeting was being done. None of the officers and managers would be hurt if they cooperated and did not resist. If there was any resistance, those would be the ones killed. All those that cooperated would be given a bonus as they were terminated from the job.

You can be sure that when word gets out on what you have done that the enforcers of the law will be coming to arrest you. You must be ready for this. When they show up have a small committee ready to invite them in so you can explain what you have done. Make this a pleasant, peaceful moment to show only cooperation and ease with the enforcers.

Have the police members leave their weapons in the room just before entering the room where you plan to meet with them. Use any excuse that will work. I will have some ideas for how to do this before you leave here.

Have your select group stand by out of site with the weapons I will send back with you. When you have explained your actions, and the conditions that led up to it, ask them to leave. The weapons they left outside the meeting will have been confiscated by your group by then, and the enforcers will now have to leave without them. Do not under any conditions return their weapons. Operate the business as though nothing had changed in what is produced there. Consider this just a trial run, and if successful will help determine the next plan of attack on the injustice of this culture," and I used a strong authoritative voice on that long speech.

The weapons that had been issued to us earlier were now used to give these "insurgents" a lesson on how they worked, and how to use them. I hoped none of them would panic, and start killing off a lot of the "guests" for no good reason.

Shortly after the guys left to go back to the factory, Speedy and I headed for the place where we had gotten the earlier weapons as I wanted to head off any attempt for more weapons to be issued as replacements to the government enforcers. This action was a small start, but if successful could be followed up with more of the same.

I knew my actions were risky and could lead to my downfall, but who cared as I had nothing to lose anyway. My life meant nothing to this planet, and since I was stuck here it meant nothing to me either. Either way they are going to know I am here, so watch out because I am a force to be reconned with. At the weapons factory, the manager was given explicit instructions that no items were to be released unless there were written instructions from me. From now on the only person that could permit release of any armaments was the King, and that is me.

Upon our arrival, Speedy was told to come in with me, and not to wait in the car anymore. I introduced him to the manager as my right-hand man and a major player in all activities. I then handed him a written order for the weapons and ammunition. Then explained once again that any requests for weapons would have to be approved by me. This was met with questioning eyes, but he knew not to argue the order. Speedy carted the load out to place in the car for the trip back while I stayed a minute to tell the manager that he was doing a good job and if it continued he would be getting a bonus in his pay. As on Earth, a good worker stays a good worker if the compensation is worth it. A title of some kind is also a good incentive so I will think of an important-sounding one for the next time to go with a perceived promotion.

On the way back Speedy suggested that I should make a trip to the district authority to meet with them over the takeover of that local factory. His suggestions on how to handle a meeting with them were a valuable addition on my efforts to deal with it. I was rapidly gaining a new admiration of the quick mind and incite that Speedy possessed on things. I would be wise to listen, and maybe even ask for his counsel on future events as it was needed.

I had been just stumbling along as events unfolded, but was now getting in so deep that there was no turning back. The moment that notification came in on the success of the takeover, and that all weapons had been confiscated with no injuries, Speedy and I headed for the district office. I showed the proper concern for the takeover and assured those authorities that proper but careful handling of this situation should be used. The case would be explained to the federal council and in due time their decision would be properly announced.

I instructed them not to do anything at this time. We will just sit back and wait to see if that insurrection will collapse because of its own inexperience or maybe even revolt by its own members. Then they may even be relieved to have the old way back again to make things run smoothly. Those district people actually seemed relieved that they did not have to take some form of drastic and possibly dangerous actions at this time. So I too was relieved that nothing would be done at this time.

With that action delayed, Speedy wasted no time getting us to the factory for a meeting with that group. When I informed them of the delay in retaliation, the next step would be to get representatives out to all the other places of businesses and factories that were operating on the same system of discrimination. Preach on the evils of discrimination used on the job and low pay. Pick out some good leaders; educate them on what to do; brag on the success of the first takeover. And of course, emphasize the bigger increase in pay on the success of having your own control. Make a point of letting them that now the King approves of the takeover, and you will have his backing on it. This was going to be a "dangerous game of chicken" but the rewards would be well worth it to gain control of your own life. That last statement seemed to be a reassuring statement on the value of taking their chances for success. I hoped they had the fortitude to give it their all.

With that decided, Speedy and I headed home. The time it took to get there gave us a chance to discuss what our next steps would be.

The only thing I had invested in this planet was my life, but that was of no use to any of these inhabitants. To them, I was an abomination. My appearance was disgusting to look at. My intellect was inferior. My speech mostly unintelligent in this language. It took their translator hanging around my neck to understand much of anything they said, so many times it had to be repeated, or I just didn't understand any of what was said so I came across as stupid in their eyes. I came to the conclusion that they would pay for this attitude, and their entire culture was going to change for the better, thanks to me. If ruthless measures had to be taken to make it happen then chalk it up to their cost in making this a better planet.

My next step would have to be the formation of an enforcement squad. The thought of somehow converting one of the soccer or one of the cleanup work groups into an armed enforcement group was tempting to try, but did not want to attempt this just yet. So gathered up some of the "inferiors" for a start. Used one of the empty lots to build a shelter house, and start their training. I had gotten pretty good by then of convincing the natives of the danger of their future if nothing was done to keep it from getting out of control in the hands of the superior class. The use of their feelings on being labeled as inferior came in handy for the argument.

I decided it was time to come up with another game that would teach strategy as well as being aggressive actions. What came to mind was the old game of hitting an opponent with a ball like they did in a gym class in high school. So I drafted some of the inferiors to make some balls and explained how they would be used. I was surprised at how fast they came up with the idea of a firm center covered with a soft material on the outside that would stick to whatever it hit.

This new game would begin with a set number of players to start. Players to be chosen by a team captain who alternated choices. Rules as to how many balls, who went first, elimination procedure, etc. would be determined by the group that designed and made the balls. The game and the name of the balls to be chosen by them as a reward for their effort.

After explaining how this game was to work this same group was asked to select a spot that had various covers to hide behind, or that would be a safe zone, etc., while the game was being played. Any player with a ball stuck to them would be immediately removed from the game, and the ball given to one of the players on the other team to be put back in play.

They also came up with the idea of having their own two teams to give demonstrations in different places on how the game worked to encourage others to get involved. I thought that was a great idea and offered to have the council pay some expenses and a stipend for them.

An aggressive group could come in handy if part of the reasoning to use force came down to intelligence versus strength. It would take a good general to change the aggression of a game to the aggression of war. It took some good manipulation of thinking that it had been superior intelligence that had put them in this inferior class, but the only thing that would correct it was strength, superior strength. With the group working as a cohesive unit and using superior strength of weapons we could succeed in righting this terrible wrong, Telling them I had access to the superior weapons was the final thing that would convinced them to get serious about a takeover.

The council had contacted Cindy since I was too busy with other thing to meet with them on plans for the king fight. She had told them to just do the same thing as the last fight. So everything was finalized with the other continent. But Cindy had strong reservations about me starting at the bottom again. This had never been done in the past, and there should be some respect shown for the reigning champion by only fighting in the last contest to be qualified for the King fight.

It was the respect I wanted, so agreed to that approach immediately. The competition on the other continent had already begun. The old champion that was the ex-King was not entered in any of the preliminaries, so knew I'd be facing a new opponent. In the meantime, it would keep me busy with those recruits training for possible combat. Word got to me that the other continent was not happy that no visits or even any kind of communique was received from the king during his reign in power. That I was a worthless King and should be eliminated to make room for the appointment of a new one. I passed this off as a nefarious attempt to get someone their champion could beat, and hence get the King back in their territory.

The fight arrangements were completed, and the other continent landed on our shore. It was a much larger contingent than usual, but nobody seemed to be concerned about it. The same procedure was followed again as the last time with a show of pride and confidence. Introduction were made and the opponent and myself were ready in the ring ready to go.

One of the opponent's helpers suddenly called for a halt. He insisted on inspecting both hand covers. He made a show of examining the two thumb gloves on the other guy, and then insisted on examining my one thumb glove. I thought that was rather funny but made no objection. At the time I didn't pay much attention to what he was looking at or doing with my gloves, so just stood there looking very superior with my larger status. I was going to make this quick to send the message I not only looked superior but would demonstrate it with a quick win.

The fight started and I was determined to wait for him to come to me. He seemed to be in no hurry to get this fight started. Seemed to me he was reluctant to get knocked on his rear, as he waved his glove close to my chin, but was way short of connecting. So I let him try a few more times without any hits just to show the fans how afraid he was to getting close enough to get hit. Showing off a bit for the audience I moved in close on him to land a few well-aimed blows. That was a mistake as those lower arms grabbed onto me and one of the top hands went right to my face. Grabbing at my eyes wasn't going to blind me, so just landed a nice sharp left that knocked him backward.

In a few moments, things began to look a bit fuzzy. The eyes were starting to burn a little. I realized then that something had been not only rubbed in my eyes but also on my gloves. This fight had to end soon or the result would be a huge loss for my position as King and everything that went with it. My hand began to itch and burn. The realization that something caustic had been put in or on my gloves, and apparently rubbed across my face registered in a now fuzzy brain. I had to get this over with in a hurry.

Catching him coming in close for another attempt at my eyes, with every ounce of strength left, a straight-on blow hit him down and out for good. Seeing that my balance was now very unsteady, the helpers rushed in to keep me on my feet. Water and a rag washed away the residue from the face, and getting the gloves jerked from my hands helped to clear up some of the confusion from the brain. My group grasped the seriousness of the situation, and quickly washed the hands, and all other exposed skin that might be contaminated.

Insisting on quick action, the entire group of those opponents and helper were taken into custody. With a fast thorough search, the vial of dope was confiscated for evidence. With the explanation that the stuff could have been fatal, that episode was the final straw in these King contests. There would be no more as long as I was on this planet. This failed assassination attempt would send a message to that other continent that their nefarious actions would not be tolerated.

The next day with the traitors still incarcerated, Speedy drove me to the Council chambers where a discussion on what to do about this foul deed was being considered. With an overnight consideration, I was going to make an example of such of deceit. So objecting to all suggestions the Council had, my idea was explained and would be carried out in public.

Old-fashioned Western justice would be done. It would be quick and cheap. A gallows would be built, a good thick rope found that I would personally tie with a hangman noose. The execution notice sent out on the time and place. I would show these people how any insurrection would be punished. I thought the Council might not approve of this punishment, so it was carried out the next morning before there could be any objections.

The bodies were loaded onto a boat, returned to the other continent with a statement of what happened, and a warning that this behavior better not happen again or I would not be so lenient with any transgressions from the inhabitants from their side of the planet.

This made me wonder how safe this job of being a king was going to be as long as it was me in that position. I had seriously considered just ending it myself, but self-preservation was hard to override. And it was a pretty cushy job even if I was totally unprepared for it.

The discrimination group came for a pow-wow on the next factory that was not willing to give one inch on changing and had fired several employees that was heading up the negotiation on it. Pushing the same idea that was used with the first factory administration and the success it had the group agreed to go with that plan.

The authorities had obviously heard of that earlier caper and brought in an armed contingent to deal with the inferiors. Several were deemed disruptors and were sentenced to be incarcerated to the Northern land. I knew what that meant for them. The big showdown was coming. This meant life or death for these inferiors, and I either had to take action, or lose the support of the inferiors, and hence the discrimination would resume. My power would be gone to get anything done here. So if it is going to take superior force then I better get the troops ready.

Speedy showed some reluctance on going back to the gun factory, but dutifully got the car ready, and got us there in good time. We had our plan to go and had even got an estimate on how many weapons we might need. The lab guy gave us a hard time on giving us what we wanted. After a few threats about what this King could do to him, he had a sudden change of attitude and complied with everything I asked for. Seems these people can be intimidated by threats just as easily as Earth people. I'll keep this in mind for future objections.

Getting back to the trainees, immediate work was started to get them ready in case a confrontation took place. These guys were preached peace and harmony, but most of them seemed anxious to get into a battle, which led me to wonder just how much this "peace" attitude had really been ingrained in them.

I needed to get some organization going with; this group. There would need to be some disconnect with me in any direct action taken on a factory as my position on it was already frowned on by the Council. I wanted to avoid trouble there as long as I could. So I called in the one person I perceived to be the most trustworthy and had the courage to get the job done. His acceptance of some praise, and being elevated to a leader confirmed I was right it his evaluation. I decided to make him a general. I didn't know the pronunciation of general in their language, so decided to just call him "General" and use that as his name. He would be in charge, but would only act on my orders. It was then left up to him to organize the rest of the group. With so darn many new names I was more than willing to use familiar names from Earth, and let them keep track of it.

Then there was the problem of finances that was becoming a big problem. Money for so many things, keeping track of where it was going, and for what, was getting too much for me. Everything from youth soccer to this ever-expanding mess. So a finance minister to keep track of it all was going to be my solution. This would then be his problem to come up with his underlings to carry out his orders to keep it running in a reasonable method. Running a country is way beyond my expertise, and makes me tired just thinking about it. No need for me to do all the work when I can have others doing it. Got to set up some cabinet-type positions to get a working government in place. Didn't know who I could trust enough to help select a person for each spot that might be necessary, so decided to turn to the ones I knew that had shown their loyalty to me. Didn't want to take them out of the job they were now doing, but give me help and advice on selecting capable and honest workers to fit into positions.

So Cindy was the first one to ask. She suggested that Blackie would be the most trustworthy in recommending a worker. I knew very little about Blackie's background so as Cindy suggested then, to just asked him about his history. But when I explained the kind of workers I needed, and the skills that would handle the job it became clear that was not what Blackie would be familiar with. So decided to go with the recommendations of the present council members and names of anybody that my circle of supporters could suggest if nothing else seemed to make sense.

I knew very little of Blackie's background, but wondering why Cindy would have such a high opinion of him I had to know more about him. So as Cindy suggested, I should just ask him to fill me in on his experience. But cookie had been here for some time as well as Blackie, so I decided to ask her about him first.

That afternoon I cornered Cookie in the dining room and explained that I wanted to appoint Blackie to a specific job and wanted to ask what she thought of his abilities to handle it. She let me know in a hurry that would depend on what the job was, and explained that she had known him for a long time, so she would give the whole story on what she knew, and to just stop her when I'd heard enough. So I'll briefly give you a shortened report on what I now know about Blackie, and hope I get out of here before any of this is discovered by the inhabitants. With their knowledge they now have of me they could easily decipher my writings, and that would put several of my cohorts in grave danger.

When Blackie was a young person he had been inspected and listed as an inferior. Along with some of his same-age friends enrolled as a group to work in the furniture factory. The work was menial and rather dirty. It didn't take much skill. You just did the part as directed, and passed it on to the next station. The work was not much of a challenge, and the pay was very poor. As a result, there was much discontent among the workers.

Complaints were handled quickly and with very severe outcomes. Often even physical punishment was used before the complainant was sent back to the job. On one such episode, Blackie had made a mistake before passing it on. When discovered that the next guy was called into the supervisor's office. He didn't blame Blackie, but said it wasn't his fault either. So Blackie was also called into the office and was sure he would be in big trouble with the boss.

As he went through the door to the office the sight of his fellow worker getting hit with fists, and being yelled at was more than his sense of fairness could take. Blackie had known these happening before and had seen the marks it had left on the victim. He picked up a chair and brought it down as hard as he could on the boss. It was obvious the guy was done for as he sank to the floor without making a sound.

in that instant, Blackie also realized that he too was done as an employee here, and would soon be incarcerated by the authorities. Once somebody was arrested and hauled in to be dealt with, they were never seen again. Knowing this, Blackie's only choice was to run. Run as far and as fast as he could before they came for him.

Knowing the force could be at his home before he could get there for any supplies, it was straight out the door of the factory and make tracks to his girlfriend's house where some supplies were quickly stuffed into a bag. His friend was not there, and with no time to write an explanation, out the door with his bag and around the house to head out thru the bush and growth of plants to vanish in the thickets that covered the ground down the hill behind the house all the way to a small creek.

He wasn't sure where he was going but needed to put some space between here and the factory. So Blackie just kept going until he came to a nice pond with some clear water. This was surrounded by bushes and shrubs of all kinds. Just up the hill, a short distance was a nice house with another smaller house next to it. He was sure a stay out of sight, and with plenty of forage, this place would do until the law gave up looking for him. That night he found a nice spot overgrown with enough foliage to offer some good protection, and keep him out of sight.

The next day with further investigation, the remains of a garden was found near the rear of the house. He could see it still had some greens in it, and some patches of fungus. It hadn't been taken care of lately, but Blackie thought with his knowledge and a little effort he could have a good enough supply of fresh food to keep him going for some time.

Cookie went on to explain that she knew nothing of travels or experiences he went thru, but one day as she was going to gather some greens at the shady side of the house there was Blackie on his hands and knees working, removing weeds, and tending to the crop of greens. He jumped to his feet to escape, but it was too late. He had been seen.

They both stood there for a moment staring at each other not knowing what to say before both demanded at the same time what each other was doing there. It took some time with half-truths and exaggerations, but finally Blackie decided he could trust cookie with his truth if she was going to permit him to stay without notifying the authorities.

It had been some time since a king had been in residence, and hence no visitors so it was almost like an abanded property. Recognizing the advantage of having Blackie continue with his garden, and expand to upkeep of the grounds around the residence, it was easy for Cookie to make it seem like he had always been there, and convince her predecessor to include him on the payroll at a small rate. She had even succeeded in moving him into the small house on the edge of the property.

 With two fugitives from the law, and with hiding in plain site where nobody thought to look, or even interest in the place to come see it then I knew here was two dependable recruits if needed. But I would like a little more information on Blackie to determine if he really would be a dependable, loyal supporter if it came down to saving any of us, or just saving yourself which is just what he had done in leaving the factory job.

A few days later I made a point of finding him out back of the house working on his garden spot of fungus. I ask him to take some time off from his fungus and join me down by the lake for some talk.

"I know about your escape from the factory. I know how you ended up here, but there are things I don't know about you, and Cookie either won't tell me, or doesn't know. She mentioned you may have had some trouble with a search party. I need to hear about that to see if your abilities will fit in with what we need to do. Tell me about that if you will." I suggested

"That is something you don't need to know after all the time I have been here doing good work keeping these grounds looking good. But I guess after all this time it could do no harm either if you will keep my story in confidence. Besides, why would you betray me now anyway", Blackie replied.

"Cookie has told you of some of it, so I will give you how I became a resident of this place. Made it to my friend's house where I stole some supplies. Then headed out of there going south because that was a rough territory with lots of hills and ravines. It would be hard for a search party to track me through the brush and brambles. Once they had given up on finding me I planned to reverse course and head north to the more comfortable territory. If they continued to hunt for me they would still be looking in the South.

I just kept working my way thru the bushes and going toward the ravine where the river ran south. I was just about there when the noise from the posse getting closer got me looking for a hiding spot with a good cover right at the top of the ravine. The noise of the river below made it hard to hear brush and stuff being moved, or even any footsteps. That night when they quit to take cover for sleep I was able to eliminate them one by one. When the light of day came then one by one they were dragged to the river to float down the river with the current. That way if they were ever found then the authorities would have no idea of where they died.

It took a few days to carefully make my way this far North. When I found this lake and the abundance of edible plants then it was an easy choice to make this the new home. I hadn't figured on getting discovered by Cookie so fast, but glad I did. This has been a good home, and I will defend it with my life as I will not run again. Cookie doesn't know everything about me or how I got to this place. You don't know it all either, and I'm not going to tell you all the details. Even if you did know, the authorities probably wouldn't believe a total outsider like you anyway."

I thanked him for sharing that much information, and agreed with him that it was probably wise not to tell too much of his escape from the factory.

That night I lay in bed pondering how to use all the information I now had about this system, my recently appointed council, and the employees at the house. This place has me confused. What I see; what is explained to me; then what I observed actually happening doesn't match. The kids I taught how to play soccer. They didn't seem to mind too much at using some muscle to convince other kids to join the team. And that big fight that took place during one game was settled, but I didn't hear of any additional admonishment for the boys involved. And that little fight with the locals when Speedie came to my rescue showed me that these people could use physical action with little concern on cultural standards as quickly and as easily as Earth people.

Somehow I have to get out of here. Suicide would do it. Maybe one last try at the space laboratory place to convince them to take me away on the next trip for fuel or water or whatever was needed. I have never been so lonely in my life. I had never thought of being alone without a friend, without a girlfriend, or even without another being of my own species to associate with or talk to, or just to stand next to. As a whole I have been treated pretty good here on this planet. So can't complain I guess. But damn it, I need someone to accept me as a real person. I know I'm just an illegal entity from other planet to these people. I should be tolerated but kept at a distance so as not to contaminate the social order of this false social place.

Received a written order from the Council today about a request from the other continent. Seems they want to restart the king competition. Their delegation will arrive in two more days. They would bring a purse of enough value to cover a good share of the expense, and to offer the present king a goodwill gift to get the competition started again. Their ambassador would like a private meeting with the King to express apologies for the past transgression and to a congenial restart. I sent word to the council that a meeting would be acceptable with their ambassador if it was held here at the King's house.

From past experience with these beings from that other continent I didn't trust any of them, so decided to make arrangements to protect myself. Had Speedie take me to where the group had been working on how to use the weapons we had gotten from the factory, and how well they were doing with training on self-defense. With the rest of that day spent on organizing how to protect me at the meeting in two days, I was satisfied that this group would have no trouble with my protection.

I didn't really expect any attack, but couldn't take anything for granted as I was still in the dark on what the other continent might consider doing to get rid of me. Maybe they did just want to re-establish a fair combat again to pick a new king, but I went over all the possibilities with my group the day before, and we were ready just in case.

The day came and my group camouflaged as servants, went with me to the Council office for a brief meeting before taking them to my house for a more informal discussion of what the rules would be for the contest, and when it would be held. The ambassador with four assistants was seated at the table with me, and eight of my group standing behind them trying to look at ease, nobody noticed that each, in turn, concentrated on his assigned target looking for anything suspicious or out of place.

With all discussion of how the match would be run, and my insistence there would be no foul play tried again, we all stood to leave. Blackie was standing behind one of them when all of a sudden he jumped forward to pin the one in front of him to the table by using the chair to hold him fast to the table. Speedie jumped in to help and took a large dagger from his lower hand. My protection squad immediately went into action, and the other three assistants plus the ambassador found themselves on the floor confined in the clutches of the rest of my men. With the look of confusion and shock at what had happened so fast and why there was no doubt that none of them knew why this had just happened.

They were all released and back on their feet except for the one that Backie and Speedie were holding. Blackie with the dagger in hand, pointed end held firmly on the man's chest while Speedie had his arm around the guy's neck with his head pulled back over the top of the chair. With nowhere to run to escape the visitors were in no position to even make any kind of a protest.

I had no experience dealing with any kind of threat such as this, but dealing with another continent this could be an opportunity to build a better relationship. Before taking them back to the Council building another meeting was in order. So seated at the table again they were going to hear my demands whether they wanted to or not.

I would not execute any of them as i originally planned after that bungled knife attack. I would expect a written apology from their authority within five days. For the King fight, these are the rules. You can only use one set of arms and hands. There will be no modification permitted in the gloves that cover the hands. I will not be a competitor in any of the preliminary bouts, nor will I compete again in a King bout. As of now, I am retired as the King, but I will stay in power until the next competition is decided. That ended my part then as they were loaded into the vehicle to go back.

I began to regret the decision to retire almost immediately, but after a good night's sleep, my feelings began to improve knowing that I still had a strong position in controlling what went on here. Pride was rapidly rebuilding in me with the thought of helping train a new fighter, and teaching skills and knowledge of what makes a good fighter. Once the word went out about it then the volunteers would be started on a training program of building strength and endurance. Skill used in being a good fighter would be taught later. My guy would not lose the next King fight.

With what I already knew from my time as a wrestler at the university the ones that would be chosen for serious training would be put to work making equipment for weight training as well as then using it. They would develop stronger muscles, and better lung capacity than they had ever had.

The attempt on my life started me thinking about what had happened to the kings and dictators of the past on Earth. Everyone I could think of had met with a violent end. Either a sneaky assignation, or a violent attack of some kind by a group that had become so mistreated by the present authority they felt the only solution was to revolt and overthrow the bums, and start a new administration. But it seemed that in most cases the deposed head was simply replaced by a new guy that gradually turned into the same thing they had just got rid of, and so it started all over again. They never seem to learn anything about their own history.

It was at this point in my time here I started to think seriously about how to end it all. I knew of the trip that got me here, and that same group should be able to take me back. They could take me back by doing the same trip as before, and replenishing the supplies they got before on that trip. Just land at the same place. A meeting with the space agency would be happening very soon.

This was my plan to get out of here. I got my home group together of Speedie, Blackie, Cindy, Cookie, and Cindy. They were all given some high-powered guns to take care of any objections. They were instructed not to use force unless it couldn't be avoided. They were informed of my Plan for getting off this planet, and back to my own. To make sure they understood why I was brought here by accident and against my will, they were told the story of how I got there. I won't repeat it here as this was explained at the beginning.

So on the very next day, we all piled into the vehicle with our weapons and a great deal of fear over confronting the high-powered space scientist, and their reaction to my demand.

We shoved open the door and crowded thru together. The guy in charge of the office was a bit rattled at suddenly being confronted with a group that were all holding weapons. He pushed on a button as he rose to meet us. A door opened and several men came thru to see what was going on but stopped immediately at seeing my group.

I quickly tried to show calmness as I told them to relax as there was nothing to fear if they did as they were told. One of them stepped forward and demanded to know what we wanted. I it made it clear that I was ready to get off this planet and go home to my own planet.

"You captured me, held me prisoner, and transported me here. I have asked several times to leave with you, but you have always just ignored it. A few days ago I demanded that you take me back when you made your next trip for supplies. The time has come. We will go as soon as your ship can be fuelled and ready to go." I said using what I thought was a good high and mighty king voice.

"You have no right nor any standing to order us to do anything," he replied. So I ended his life right there with a shot to the chest. The rest of my gang immediately raised their weapons but did not fire.

They could tell we had the upper hand so stood frozen in place as I repeated some of my demands adding that they had ten days to get the ship ready to go. Then decided to make it as forceful as I could.

"If your crew and ship is not ready to go in ten days I, my guards, and my entire army will kill every one of you, and blow up this entire compound."

We turned and stomped out without saying another word. Ounce outside I could see all the questionable looks, so on the way back explained that maybe we didn't need an army or really no force except just this little group. But they didn't know that so hopefully it scared them enough to comply.

The next three days were spent discussing how they should deal with their country now with me out of the picture. I thought about making suggestions on how to shape their new future without me but decided to hell with it. They can figure it out themselves. I've had enough of this place and I know they have had enough of me, so at least I'll try to get out of here with a few friends.

The next day I wanted to know how the space guys were doing with preparations to leave. Thought better about going myself and taking a chance on getting assigned. That would give them an easy solution getting rid of me. Speedy and Blackie volunteered to go check on it for me.

On their return the report was good. They should be ready to go. Speedy told me there was even a special capsule for me to ride in down to the surface. It would be released over my area, and guided down by signals from the big ship to a smooth landing on Earth as close as possible to the spot where I was originally captured.

I will be ready to get home. I plan to travel the world giving speeches on what life was like on this planet. Two more days and I'm out of here. I don't care if Cindy thinks I'm ugly. I'm going to kiss her goodbye right on

I AM THE ALIEN

her lips anyway.

I have become so confused on what is developing on this continent, and how to guide it, that it will be the best for both this place and me to get off this ride now.

I don't want to be assigned here, and I'm so homesick for my home and country. I go see my capsule for the ride down to Earth for a fit, and instructions. So this is it from here, and I'll be smiling as I greet you on my return.

ABOUT THE AUTHOR

The author was born and raised in the farm country of Iowa. After graduating from the University of Iowa he spent forty years as a mathematics teacher in Iowa and Texas. The next few years were his home on the bluffs in Wisconsin where he and Carol enjoyed watching the eagles fly by on their migration, and the barges, riverboats, and fishermen on the Mississippi River below. They are now back to their native state of Iowa enjoying retirement telling tall stories for the grandchildren. This story **"I AM THE ALIEN"** is his last tall tale.

www.ingramcontent.com/pod-product-compliance
Lightning Source LLC
Chambersburg PA
CBHW052359220526
45465CB00003BB/1165